DIVERSITY IN MOUNTAIN SYSTEMS

COLLOQUIUM GEOGRAPHICUM
Band 31 ISSN 0588-3253

Herausgeber • *Editor*
Geographisches Institut der Universität Bonn
Department of Geography, University of Bonn

Verantwortlicher Herausgeber • *Editor-in-Chief*   W. Schenk
Schriftleitung • *Editorial Management*   A. Lunkenheimer

ASGARD-VERLAG SANKT AUGUSTIN 2008

Diversity in Mountain Systems

Studies in Mountain Environments: prepared by
members of the 'Working Group on High Mountains',
German Geographical Society (DGfG). Submitted to the
Commission C04.08, 'Diversity in Mountain Systems',
of the International Geographical Union (IGU) on the
occasion of the 31st International Geographical
Congress, Tunis 2008

Edited by
Jörg LÖFFLER and JÖRG STADELBAUER

With articles by

Friederike Grüninger & Thomas Fickert,
Hermann Kreutzmann, Matthias Kuhle,
Jörg Löffler & Roland Pape,
Marcus Nüsser & Martin Gerwin,
Perdita Pohle, Matthias Schmidt,
Andrea Schneider & Jörg Stadelbauer,
Christoph Stadel

In Kommission bei • *on consignment by*
Asgard-Verlag • Sankt Augustin

alle Rechte vorbehalten
all rights reserved

Bibliographische Information der Deutschen Bibliothek:
Die Deutsche Bibliothek verzeichnet diese Publikation in der
Deutschen Nationalbibliographie; detaillierte bibliographische
Daten sind unter http://dnb.ddb.de abrufbar.

ISBN: 978-3-537-87431-3

© 2008 Asgard-Verlag Dr. Werner Hippe GmbH, 53757 Sankt Augustin
Druck • *Print*   Druckerei Martin Roesberg, 53347 Alfter-Impekoven

Contents

Diversity in Mountain Systems – An Introduction
Jörg Stadelbauer — 7

Geomorphological Altitudinal Levels of the Mountains and the Influence of the Relief on the Glacier Altitudinal Level During the Quaternary Climate Change as Basis of a Glacier Typology
Matthias Kuhle — 21

Diversity is in the Eye of the Beholder – Plant Diversity Patterns and their Relation to Scale: Examples from the South-Western USA
Friederike Grüninger & Thomas Fickert — 33

Plant Diversity Patterns and Reindeer Pastoralism in Northern Norwegian Mountain Systems
Jörg Löffler & Roland Pape — 57

Agrarian Diversity, Resilience and Adaptation of Andean Agriculture and Rural Communities
Christoph Stadel — 73

Indigenous and Local Concepts of Land Use and Biodiversity Management in the Andes of Southern Ecuador
Perdita Pohle — 89

Diversity, Complexity and Dynamics: Land Use Patterns in the Central Himalayas of Kumaon, Northern India
Marcus Nüsser & Martin Gerwin — 107

Diversity in Mountain Tourism: the Example of Kyrgyzstan
Andrea Schneider & Jörg Stadelbauer — 121

Political Ecology in High Mountains: the Web of Actors, Interests and Institutions in Kyrgyzstan's Mountains
Matthias Schmidt — 139

Boundary-Making and Geopolitical Diversity in the Pamirian Knot
Hermann Kreutzmann — 155

Diversity in Mountain Systems – An Introduction

Jörg Stadelbauer[1]

International mountain research is generally traced back to Alexander von Humboldt and his expedition to South America. The achievements of this famous natural scientist, who always went beyond natural scientific methods of measuring and who took the social circumstances in the countries visited into account, were still appreciated among alpine geographers of the 19th and at the turn of the 20th century, for example by the Schlagintweit brothers or Gottfried Merzbacher (Merzbacher 1904, 1907), as well as by geomorphologists of the early 20th century. The model representations regarding the cycle of erosion, developed by William Morris Davis, need to be mentioned also and the model of the glacial series, which goes back to Penck and Brückner.

The wars of the 20th century expanded military activities to the mountains and generated the emergence of new research disciplines. Aerial photography and photogrammetry were closely linked to the research of high mountains, as early works of Carl Troll indicate (Pillewizer 1986; Troll 1966a). Since Troll applied an ecological perspective (geo-ecological approach), he became one of the crucial initiators of international (high) mountain research after the Second World War (Troll 1966). Therewith, he also paved the way for international research programs such as Man and Biosphere, which have been carried out since the 1970s, increasingly shifting the focus towards human-environmental interactions.[2]

The high mountain research experienced a significant upturn, in particular during the 1990s, after UNCED announced sustainability as the paradigm for development politics in Rio de Janeiro 1992, explicitly embracing mountain areas. Chapter 13 – mainly formulated by Jack Ives and Bruno Messerli – of the World Conference Agenda, became a guideline for at least one decade. Two publications, which provided summaries and global overviews, contributed to the promotion of mountain areas for a large public – scientifically, as well as actor-oriented (Stone 1992; Messerli & Ives 1997). A political peak was reached shortly after the turn of the last century, when the UN announced that 2002 would be the International Year of Mountains, which resulted in the Global Mountain Summit in Bishkek during the autumn (Kyrgyzstan) (Ives & Messerli 2001). In the meantime, information and research networks had been established, in which Mountain Forum and Mountain Research Initiative was emphasized. Not much has happened in mountain research since 2002. However, impulses emerged and currently, research activities are conducted in almost all mountain areas around the world. In this context, collected papers, which provided overviews and comparisons, can be regarded as most influential (Stone 1992; Ives & Messerli 1997; Burga; Klötzli & Grabherr 2004).

[1] University of Freiburg, Department of Human Geography, D-79085 Freiburg; joerg.stadelbauer@geographie.uni-freiburg.de.
[2] The project 6 of the UNESCO-programme „Man and Biosphere" was carried out under the indicative title of „Human Impacts on Mountain Ecosystems". A very broad overview on the history of high mountain research is given by Uhlig and Haffner (1984a) in their introduction to their edited collection of papers, which has a strong orientation towards the history of the discipline (Uhlig & Haffner 1984).

Mountain research within IGU has realigned along mainstreams of the international research sphere regarding the discipline, as well as the debate on development questions. The implication has become manifest in multiple paradigm shifts, which will provide a focus for the following paragraphs. This retrospect contributes to substantiating the concept of diversity in this volume and integrate it into the discipline's historical context.

Mountain research in the context of IGU, facing the change of research paradigms

Three-dimensionality of mountain areas – Alexander von Humboldt's legacy

An IGU Commission on Mountains was first established in 1968 on the occasion of the 12th IGU General Assembly in New Delhi, suggested by Carl Troll and founded under the name "Commission on High-Altitude Geoecology". Previously, efforts had been made that were linked to Troll's IGU presidency (1960-1964). In 1966, he initiated a joint preparatory symposium with UNESCO at the IGU regional conference in Mexico. However, he was not able to participate due to health reasons (TROLL 1968). The impulse to create the new commission came at a time, when the number of IGU Commissions was still small. In particular, the major sub-disciplines were trying to get a chance to become noticeable. In contrast to most commissions dealing with factual issues, the first High Mountains Commission aimed for an integrative-landscape ecological approach. The dominant paradigm in this was the three-dimensionality of mountain areas. It drew on research and illustrations by Alexander von Humboldt, was further influenced by Lautensach's "*Formenwandellehre*", and received further decisive impetus from Troll's own works (TROLL 1941, 1959, 1962, 1975). Besides definitional questions regarding the terminology of low range mountains and high mountains (SCHWEIZER 1984), particularly bio-geographical analyses were at the center of interest. Yet the differentiation of geomorphological processes and forms, derived from climatic morphology, is also based on the paradigm of three-dimensionality (HÖLLERMANN 1976). Concerning the human geographic analysis, UHLIG (1984) took up this idea in order to develop the concept of vertical stratification of utilization stages and interdependencies between them.

High mountain regions and high latitude zones – a questionable analogy

The idea of analogy and comparison was introduced by Troll – following the intellectual path of Alexander von Humboldt – via studies dealing with individual phenomena ("*Büsserschnee*", solifluction, tree ferns), which appeared characteristically for tropical and sub-tropical mountain areas. According to the landscape ecological approach, there also was a need to apply it to the analogy between altitude and geographical latitude. While the scientific journal *Arctic, Antarctic and Alpine Research* (formerly *Arctic and Alpine Research*) was and is able to combine those perspectives, it turned out that a direct comparison of high mountains with high latitude zones was not feasible, because other process cycles had to be analyzed – even though outer appearances were similar. Therefore, the analogy was not taken up by the Mountain Commission, but rather by the Commission on Rural Development, managed by György Enyedi, which was temporarily under significant influence of Uuno Varjo and resulted in the creation of a Sub-Commission on Rural Development in Highlands and High-latitude Zones that particularly addressed developmental problems of the circumpolar region (Majoral & Lopez 1983; Leidlmair & Frantz 1985; Varjo 1985).

As Humboldt and Troll promoted the comparison of mountain areas in detailed analyses of vegetation, the analogy approach encouraged comparative analysis once again. The comparison of high mountains, which was exemplarily highlighted in context of natural geographical phenomena by Rathjens (1966, 1982), was then applied in human geography by Grötzbach (1975). As a matter of fact, societal aspects were not as prominently represented among the member of the Commission and their activities yet. However, the renaming into "Mountain Geoecology" in 1976 allows a certain degree of flexibility, which is inherent in the term "geoecology", and accounts for this type of open approach.

Human-environmental interactions, in high mountains und lowlands

The focus on high mountains and (unsettled) areas of the high latitude zones as well as landscape ecological research approaches implicated that the Commission initially concentrated on physical-geographical topics. Due to the recognition of ethnical-cultural differentiation at the North-Caucasus-Symposium in 1976, the human-geographical aspects moved further towards the centre of attention. However, in 1980 at the symposium in Tsukuba only very few talks addressed issues, which had a predominantly human-geographical orientation. Nevertheless, a new paradigm shift, which became apparent in 1984 at the latest, was prepared on the occasion of an excursion to the Alps, which was connected to the Geography Conference in Paris: the interaction between humans and environment. The activities of the UNESCO-Programme Man and Biosphere (MAB), whose 6th program sector was concerned with mountain areas, served as preparation (UNESCO 1974). The results of four test areas in Switzerland were already partly available at this point in time; they provided the base for a collection of papers (Brugger et al. 1984). The data was summarized in a book published by Paul Messerli (1989) (see also Price 1995 with its comprehensive appraisal of MAB-articles). In German geography, Uhlig (1984), based on the tradition of vertical spatial differentiation, emphasized interaction between geo-ecological and human-geographical variables. This gave higher priority to the examination of mountain areas as a whole and replaced the strong focus on unsettled areas of high mountains. The idea of a systemic interrelation between mountains and mountain forelands went a step further. An approach of applied mountain research in the tropics and subtropics was derived was this in cooperation with the United Nations University – initially using the example of Southeast Asia (Ives 1980). The analysis of correlations between spatially differentiated development of the Himalayas and around the Ganges-Brahmaputra lowlands is considered the basis for this approach, whereas Ives and Messerli (1989) reject the deterministic explanation that the flooding of the Ganges-Delta area in Bangladesh happens due to logging in mountain areas. In a broader sense, high mountains were increasingly perceived as resources for the lowlands (Rathjens 1982; Winiger 1992: 406). A conference, which was organized by the IGU-Commission in cooperation with the Academy of Sciences in 1983 in Mainz, highlighted the human-environmental relation, referring almost solely to land use (LAUER 1984). Moreover, mineral ores and water resources need to be considered likewise as interdependencies of mountain nomadism and pasturing or scenic resources for tourism use. After all, the significance of anthropogenic variables, which cause current problems in high mountains were strikingly demonstrated in case studies by ALLAN, KNAPP and STADEL (1988), edited in a collection of papers.[3]

The increasing differentiation in questions of mountain research, the nexus of natural and social sciences approaches, and the orientation towards implementation within the frame of development policy required a certain degree of institutionalization. The journal Mountain Research and Development has been published since 1981 and represents the most popular forum worldwide. With its two major categories, "Research" and "Development", the journal facilitates cooperation between research and practice. Furthermore, the International Centre for Integrated Mountain Development (ICIMOD) was founded in 1983 in Kathmandu. The foundation of the International Mountain Society was followed by the foundation of the African Mountain Association (1986) and the Andean Mountain Association (1991).

Due to the fact that the Mountain Commission also dealt with questions of resource allocation, another renaming seemed necessary in 1988. The Commission was named "Mountain Geoecology and Resource Development" until 1992, which reflected its new purpose.

Sustainable development in mountain areas – a challenge to science and politics

An important milestone was marked by the UN-Conference on Environment and Development in Rio de Janeiro 1992, which declared the term 'sustainability', introduced by the so-called Brundtland-Report in 1987, a paradigm for development policy and lifted it up to a political level. This gave a new impetus to the IGU-Mountain Commission. A symposium took place, partly in Staufen near Freiburg in 1994 (including an excursion to the Black forest) and partly in the Hohen Tauern and in the Berchtesgarden National Park, which was supposed to highlight the existence of specific problems and development processes in low range mountains.[4] Whereas Troll explicitly referred to the high mountains area, now all mountain areas should be considered – responding to the review of 1992 (STONE 1992; An Appeal for the Mountains). Thereby, the paradigm of sustainability was given more and more priority.

Consequentially, the IGU-Mountain Commission obtained the new name "Mountain geoecology and sustainable development" at a symposium in Davis, CA. that preceded the main conference in Washington D.C. Thus, after the resource aspect had been raised, a development objective was also raised, which facilitated not only the realization of basic but also applied research. However, KREUTZMANN (2001) emphasized that there are still knowledge deficits regarding developmental problems of high mountain regions.

[3] In German-speaking geography, Erwin GRÖTZBACH (1975, 1982) formulated several basic considerations on Human Geography in high mountains and then jointly organized a symposium with Gisbert Rinschede in 1983, of which the outcome has been published (GRÖTZBACH & RINSCHEDE 1984).
[4] Some of the articles were published in a topic-specific volume of Mountain Research and Development vol. 15, no. 3, 1995.

The high risk potential in particular is among the problems, which oppose sustainable development in mountain regions. The seismicity of numerous high mountains, movements of earth masses, slope erosion, disasters caused by the run-off of mountain rivers, and other phenomena are relevant in this context, which once again points to human-environmental relations: On the one hand, montane societies need to react to challenges of the mountain environment, on the other hand, they often increase the risk of natural disasters due to utilization and overexploitation of resources (Ives & Messerli 1984; Messerli & Hofer 1992). Whereas natural threats to society can hardly be eliminated, anthropogenic intervention requires an adequate response from society. In order to protect mountain regions, institutions such as CIPRA (*Commission Internationale pour la Protection des Alpes*) have been established.

Diversity and systems analysis – refinement of research instruments

The broad overviews of 1992 and 1997, which summarized the acquired knowledge on high mountains and corresponding problems, but also the increasing challenges for governance, revealed that our knowledge about mountain areas is still insufficient. New methodological approaches evolved in the 1990s, in both the physical and human-geographical fields and were aimed at exact measurements and the inclusion of individual perceptions and actions (Winiger & Bendix 2000). Thereby, the variety of mountain areas moved more into the foreground. Consequentially, the new key term became diversity in 2000. The term was not meant to be reduced to biodiversity, but to demonstrate the evolving variety in all fields, because mountains need to be looked at in accordance to their particular regional level of scale. Hence, the paradigm of diversity also acts, in a way, as an alternative draft to the globalization process and therefore corresponds to the consistently highlighted diversity of mountain areas.

The biodiversity of high mountains received specific recognition in this context. Hence, the international research community was able to set up an observation network in several biosphere-reserves that was primarily designed to keep track of the impacts of global change upon mountain areas. In contrast to naturally originated changes of basic conditions, anthropogenic developments were rather neglected (Huber et al. 2005; Research Programme Change in Mountain Regions – GLOCHAMORE). At this point, the necessity for further research at the interface between approaches of natural and social sciences became apparent.

International mountain research – contributions from national research initiatives

The focus of IGU Commissions has always been internationally oriented. The point is to present the numerous individual research projects, to contrast them and reunite their results. The variety of theoretical and methodological approaches and the diversity of working strategies – which in many cases are adjusted to the (research) object – facilitate the additional benefits that can be achieved through collaboration.

Furthermore, the international orientation is a stimulus for the work at the national level. Numerous member states of IGU have working groups in place, in which particular research activities in the mountains are clustered. Thus for example, there is a Mountain Specialty Group within the framework of the Association of American Geographers as well as within the Royal Geographical Society. In the USA, the Mountain Institute in Washington has set up an international research network, whose main emphasis is on the Appalachians, the Andes and the Himalayas, while the Rocky Mountain Institute (Snowmass and Boulder, Colorado) focuses on applied research with regard to the use of resources. In Canada, the Banff Centre should be mentioned. The Institute of Mountain Hazards and Environment of the Chinese Academy of Sciences at Chengdu takes care of this task in the People's Republic of China jointly with research institutes. Within the former Soviet Union and in Russia, the Geographical Institute of the Academy of Sciences ("Institut geografii Akademii nauk SSSR" resp. "*Institut geografii Rossiiskoi Akademii nauk*" [IGRAN]) took care of this kind of coordination.

The same spatial and institutional concentration of mountain research towards certain institutions can be detected in other countries. For example, in France, the "*Institut de Géographie Alpine*" (Grenoble) has conducted mountain research in alpine regions for several decades and publishes the professional journal "Revue des Géographie Alpine"; while the CERAMAC (Centre d'Etudes et de Recherches Appliquées au Massif Central, à la moyenne montagne et aux espaces fragiles), which is located in Clermont-Ferrand, engages in analysis of low range mountains and interrelated problems (*Moyennes montagnes européennes* 1999; *Crises et mutations* … 2003). In Switzerland, the Geographical Institute of the University of Bern leads the way. Their work can be traced back at the least to the 1970s and the successful implementation of research in the context of the Man and Biosphere Program.

In the German-speaking area in Central Europe (Germany, Austria, Switzerland) several approaches of institutionalization can be referred to. On the base of research conducted in Nepal, and on the publication of detailed topographical maps, a working committee for comparative high altitude mountain research was already established in 1965 and was interdisciplinary oriented. In 2001, the Bavarian Academy of Sciences in Munich assigned a scientific committee for mountain research with institutes in Grenoble, Bern, Rome, Vienna/Innsbruck, and Ljubljana, for research mainly directed towards the alpine region. The committee is also involved in activities of the International Committee for Alpine Research (ISCAR) and scientifically accompanies the activities of CIPRA. Within the German Society for Geography two research groups deal with mountain research, one, founded in 1994, is called "*Arbeitskreis Geoökologie der Hochgebirge*", since 2004 "*Arbeitskreis Hochgebirge*" and the "*Arbeitskreis für vergleichende Mittelgebirgsforschung*", whose main focus is low range mountain areas, which have been exposed to anthropogenic transformation processes for several centuries. However, it predominantly works geomorphologically. The high significance of mountain research in Germany can only be explained by the initiative of several academic researchers, whose research activities were mainly dedicated towards mountains. Besides the people involved in IGU (Carl Troll, Peter Höllermann, Matthias Winiger) particularly Harald Uhlig, Carl Rathjens, Wilhelm Lauer and Helmut Heuberger should be mentioned.

This tradition is carried on by the current generation, whereas personal engagement and individual interest play a central role here. Some people were able to contribute to this volume and are not mentioned here. Others were kept occupied by other tasks, so that they could not provide an article, but are listed here due to their extraordinary significance for mountain research: Michael Richter examines aspects of biodiversity connected to the analysis of three-dimensional differentiation of mountain vegetation; Georg Miehe investigates current developments of the Himalayas and Tibet in order to draw conclusions about historical dynamics of vegetation and landscape. Willibald Haffner, Ulrike Müller-Böker and Perdita Pohle dedicated their attention especially to the Nepal Himalayas and combined thereby the landscape ecological approach with a socio-economic and also a cultural-historical one. Furthermore, Friedrich-Karl Holtmeier deals with the alpine and sub-arctic region, and Jörg Bendix contributes to new findings on the three-dimensionality of climatic conditions.

Besides the Man and Biosphere projects, other major programs, which were financed by the German Research Foundation, are also important. They targeted integrating socio-scientific questions into mountain research, namely the programs "*Siedlungsprozesse und Staatenbildungen im Tibetischem Himalaya*" (Settlement processes and state-building in the Tibetan Himalaya) and "*Kulturraum Karakorum*" (Cultural Area Karakoram).

Diversity

The name of the commission and the title of the publication link the diversity paradigm, which was initially developed in biogeography and then was taken up in ethnogeography and humanities, with a systems approach. It can generally be traced back to Carl Troll, for whom the term geoecology already implied an interaction of different elements. Adding to that, experience was gained via MAB-Projects, because for both, Austria (Ötztal/Obergurgl) and Switzerland, model representations based on systemic correlations were developed. Later on, the Swiss model was taken up and modified by Bruno Messerli several times, in order to highlight, in particular, its explanatory power with regard to land use and land use change, which represents an interface between the influence of geo-factors of natural origin and the societal-political-economic framework. In addition, spatial interaction and other systems elements have been drawn upon in numerous other mountain research activities (e. g. highland-lowland interactive systems; cf. Ives 1980).

The significance of mountains as "water towers", as regions of high biodiversity, as well as of high ethnic and cultural diversity, the fragility of natural ecosystems exposed to human impact, the continuous adaptation of mountain civilizations to the demands of metropolitan vicinities, as well as to the challenges of modernisation; and, finally, influences of global climate change are exemplarily mentioned (*Mountain Agenda* 1998).

A summarizing basis could be seen in several topics and was pointed out in the Bishkek Mountain Platform which had been prepared for the Bishkek Mountain Summit (28 October through 2 November 2002 in Bishkek, capital of Kyrgyzstan):[5]

> "Mountain areas cover 24.6 percent of the Earth's land surface and host 12 percent of its people. Mountains provide vital resources for both mountain and lowland people, including fresh water for at least half of humanity, critical reserves of biodiversity, food, forests and minerals. They are culturally rich and provide places for the physical and spiritual recreation of the inhabitants of our increasingly urbanized planet. The

[5] http://www.ourplanet.com/imgversn/134/bishkek.html (03/06/2008).

people of mountain areas face major challenges. About half of the world's approximately 700 million mountain inhabitants are vulnerable to food shortages and chronic malnutrition. Mountain people, particularly disadvantaged groups such as women and children, suffer more than others from the unequal distribution of assets and from conflicts. Policy decisions influencing the use of mountain resources are generally made in centres of power far from mountain communities, which are often politically marginalized and receive inadequate compensation for mountain resources, services and products. Mountain ecosystems are exceedingly diverse but fragile because of their steep slopes, altitude and extreme landscapes. Many of these ecosystems are being degraded because farmers are forced to apply unsustainable agricultural practices and by inappropriate development. Climate change, natural hazards and other forces also threaten the complex webs of life that mountains support. The consequences of poverty and environmental degradation reach far beyond mountain communities, through war, terrorism, refugee movements, migration, loss of human potential, drought, famine, and escalating numbers of landslides, mudslides, catastrophic floods and other natural disasters in highlands and lowlands. Moreover, the rapid melting of mountain glaciers and degradation of watersheds is reducing the availability of life-sustaining water and increasing the potential of conflict over dwindling supplies."

Whereas models are usually meant to be universally applicable, in this case they are confronted by multiple appearances, which can barely be classified. Each mountain area has its peculiar characteristics. The two descriptions of STONE (1992) and MESSERLI & IVES (1997), which provide an overview, have made it clear that our knowledge on regional specifications is still insufficient and therefore generalizations should be avoided – particularly regarding the human-environmental context. Even widely known works (BURGA et al. 2004; DECH et al. 2005) are only able to fill the gap incompletely. At the same time, the research community has made progress in detecting interrelations between phenomena during the last decades, thanks to improved accuracy of research instruments.

A special relevance in this systems context can be assigned to the discussion on mountain research as a kind of research interface of science and humanities. We have always known two branches of geography, of which one is based on approaches in natural sciences and one in social sciences. After having observed rather increasing discrepancies between the two sub-disciplines for the past decades and the symptomatic lack of communication, the last years have been dominated by the emergence of a "third pillar", which try to unite the two aspects. Mountain research appears to be particularly suitable for this. Consequentially, a number of the following case studies are intentionally located at the interface.

Primarily, the presented collection of papers is based on three deliberations:
- A systems approach takes center stage because it represents a central paradigm in geography: Spatial appearances and processes are interlinked and influence and determine each other. The debate as to whether natural or anthropogenic factors are dominant and controlling or whether both coincide ("interface") is obsolete. Appearances and processes of mountain areas are respectively perceived as systems, even though there might be cases in which it plays a minor role, whether the whole

system (which is often not possible) is dealt with or only one part of it is analyzed. At all times, it is essential to identify the driving forces for systemic processes and changes. Depending on the particular driving force systems, they can even possibly be divided into groups – I am reluctant to call it "classified" because this seems too schematic, instead the word "clustered" would perhaps be more appropriate.

- A second way of grouping result from "diversity" itself. It is very common to talk about biodiversity, whereas we are also aware of the existence of "ethno-diversity", "relief-diversity" etc. Each subdiscipline of geography contributes to the compilation of spatial diversities. Linking this thought to the former one, the objective is to recognize and capture the variety of appearances of mountains, as such, and at the same time, to realize that biodiversity is not exclusively related to the bios, but also to driving forces such as politics of land use change, relief-dynamics or economic demands, which have to be considered. It also obtains a political dimension itself, if the thought of protection becomes embraced. Therefore, an investigation of biodiversity in mountains needs to look beyond biospheres as such, likewise ethno-diversity cannot only be concerned with cultural diversity in mountain areas, but is also asked to include the appraisal of natural resources, reflected in the economic situation, as well as regulating acculturation processes or dynamics of settlement formation.

- When merging these two thoughts, while keeping larger mountain areas in mind, the risk of superficiality is high. Although indeed many instrumental factors and phenomena are specified, mutual influences are not demonstrated adequately. Therefore, it seems reasonable to present exemplary case studies, which of course cannot embrace an overall picture of diversity in mountain areas, but still and at the least indicate this variety. Hence, the intended volume benefits from a wider variety of examples that are subject to the same basic concept. Thus, the examples, in which theoretical transfers to other spatial systems are feasible and which should always be looked at in the larger context, need to take center stage.

Perspective: Mountain areas under the influence of global change

The new methods of measurement have demonstrated the sensitivity of mountain-ecosystems; a network of small-scale analyses, primarily in biosphere-reserves, indicates the current dynamics (GLOCHAMORE). At the same time, research deficits can be detected concerning impacts of global changes on society and land use, as well as the contributions of the montane society to global change. Therefore, it becomes more than obvious that future mountain research should be given a much stronger orientation towards the global change impact on mountain areas and corresponding responses from montane systems: A new IGU Commission "*Mountain response to global change*" could put emphasis on this intention and initiate the necessary international research activities in this context. Thereby, the fact needs to be considered that reflections on global change should not solely be limited to climate change and resulting environmental transformations, but also include other aspects. Among these aspects is economic globalization, which causes social changes that evolve from rising disparities between regions, social stratum and the differentiated access to resources, but also due to divergent population dynamics in more and less developed countries that result in inequalities.

Among environmental changes the rapid melting of mountain glaciers deserves particular attention, because it is closely connected to the destabilization of slopes that were so far underlying conditions of permafrost, but now are mobilized and therefore pose a latent threat to residential areas and economic activities. Environmental hazards, which tend to increase, and the growing extent of natural disasters, are additional side-effects of global change, which particularly affect mountain areas (DIKAU et al. 2004).

Postscript

The publishing of this volume on the topic of diversity of mountain areas, especially in the renowned *Colloquium Geographicum* edited by the Geographical Institute of the University Bonn, cannot take place without remembering the decade-long tradition of mountain research in Bonn. Forty years ago a collection of papers was released in this series that summarized the results of the high mountains conference in Mexico in 1966 that had been carried out in the preparation of founding the first IGU Mountain Commission. Troll's successor Wilhelm Lauer (1923 – 2007) would have been 85 years old in the spring of 2008, his academic "grandson" who holds the chair, Matthias Winiger, celebrated his 65th birthday in 2008. And we want to mention more than just those milestone birthdays. Peter Höllermann (born in 1931), another champion in international mountain research, must be referred to here and likewise Jack D. Ives and Bruno Messerli, both honorary members of the Commission 'Diversity in Mountain Systems' who all turn 77 in 2008. Sincere thanks to all of them for their commitment to maintaining the continuity of mountain research.

References

ALLAN, Nigel J. R. (1986): Accessibility and altitudinal zonation models of mountains. Mountain Research and Development 6 (3):185–194.
ALLAN, Nigel J. R., Gregory W. KNAPP, Christoph STADEL [eds] (1988): Human Impact on Mountains. Totowa.
An Appeal for the Mountains. Prepared on the occasion of the United Nations Conference on Environment and Development (UNCED), Rio de Janeiro, June 1992 (1992). Bern.
BRUGGER, E.A., G. FURRER, Bruno MESSERLI, Paul MESSERLI [eds] (1984): Umbruch im Berggebiet: Die Entwicklung des schweizerischen Berggebiets zwischen Eigenständigkeit und Abhängigkeit aus ökonomischer und ökologischer Sicht. Bern, Stuttgart.
BURGA, Conradin A., Frank KLÖTZLI, Georg GRABHERR [eds] (2004): Gebirge der Erde: Landschaft, Klima, Pflanzenwelt. Stuttgart.
Crises et mutations des agricultures de montagne. Colloque international en hommage au Professeur Christian Mignon (2003). Clermont-Ferrand (CERAMAC; 20).
DECH, S., R. MESSNER, R. GLASER, R.-P. MÄRTIN (2005): Berge aus dem All. München.
DIKAU, Richard, KREUTZMANN, Hermann & Matthias WINIGER (2002): Zwischen Alpen, Anden und Himalaya. In: EHLERS, E. & H. LESER [eds]: Geographie heute – für die Welt von morgen. Gotha. 82–89.
EHLERS, Eckart, Hermann KREUTZMANN [eds] (2000): High mountain pastoralism in Northern Pakistan. Erdkundliches Wissen 132. Stuttgart.
FORSYTH, T. (1998): Mountain myths revisited: Integrating natural and social environmental science. Mountain Research and Development 18 (2):107–116.

Grötzbach, Erwin (1975): Überlegungen zu einer vergleichenden Kulturgeographie altweltlicher Hochgebirge. In: 40. Deutscher Geographentag Innsbruck 19. bis 25. Mai 1975. Tagungsbericht und wissenschaftliche Abhandlungen. Wiesbaden 1976: 109-118; wieder abgedruckt in: Uhlig & Haffner [eds] (1984): 480–491.

Grötzbach, Erwin (1982): Das Hochgebirge als menschlicher Lebensraum. Eichstätter Hochschulreden 33. München.

Grötzbach, Erwin, Gisbert Rinschede [eds] (1984): Beiträge zur vergleichenden Kulturgeographie der Hochgebirge. Regensburg (= Eichstätter Beiträge; 12).

Henning, Ingrid (1970): Bericht über das internationale Symposium der IGU Commission on High-Altitude Geoecology über die Landschaftsökologie der Hochgebirge Eurasiens. Erdkunde 24: 234–236.

Hewitt, Ken (1992): Mountain hazards. GeoJournal 27 (1):47–60.

Höllermann, Peter (1976): Probleme der rezenten geomorphologischen Höhenstufung im Rahmen einer vergleichenden Hochgebirgsgeographie. In: 40. Deutscher Geographentag Innsbruck 19. bis 25. Mai 1975. Tagungsbericht und wissenschaftliche Abhandlungen. Wiesbaden 1976: 61–75.

Huber, U. M., H. K. M. Bugmann, M. A. Reasoner [eds] (2005): Global change and mountain regions: An overview of current knowledge. Dordrecht.

Ives, Jack D. (1980): Highland-lowland Interactive Systems in the Humid Tropics and Subtropics: the need for a conceptual basis for an applied research programme. In: Ives, Jack, D., Sanga Sabhasri, Pisit Voraurai [eds]: Conservation and Development in Northern Thailand. Proceedings of a Programmatic Workshop … Tokyo: 3-8.

Ives, Jack D., Bruno Messerli (1984): Stability and instability of mountain ecosystems: lessons learned and recommendation for the future. Mountain Research and Development 4 (1): 63–71.

Ives, Jack D., Bruno Messerli (1989): The Himalayan Dilemma. London.

Ives, Jack D., Bruno Messerli (1990): Progress in Theoretical and Applied Mountain Research, 1973-1989, and Major Future Needs. In: Mountain Research and Development 10: 101–127.

Ives, Jack D., Bruno Messerli (2001): Perspektiven für die zukünftige Gebirgsforschung und Gebirgsentwicklung. In: Geographische Rundschau 53 (12): 4–7.

Kreutzmann, Hermann (1993): Entwicklungstendenzen in den Hochgebirgsregionen des indischen Subkontinents. Die Erde 124 (1):1–18.

Kreutzmann, Hermann (2001): Development indicators for mountain regions. Mountain Research and Development 21 (2):34–41.

Kreutzmann, Hermann (2001): Entwicklungsforschung und Hochgebirge. In: Geographische Rundschau 53 (12): 8–15.

Lauer, Wilhelm [ed.] (1984): Natural Environment and Man in Tropical Mountain Ecosystems. Natur und Mensch im Ökosystem tropischer Hochgebirge. Stuttgart (Erdwissenschaftliche Forschung; 18).

Leidlmair, Adolf, Klaus Frantz (1985): Environment and Human Life in Highlands and High-Latitude Zones. Proceedings of a Symposium … Innsbruck (= Innsbrucker Geographische Studien; 13).

Majoral, Roser, Francesc Lopez [eds] (1983): Rural Life and the Exploitation of Natural resources in Highlands and High-Latitude Zones. Proceedings of a Symposium … Barcelona.

MERZBACHER, Gottfried (1904): Vorläufiger Bericht über eine in den Jahren 1902 und 1903 ausgeführte Forschungsreise in den zentralen Tian-Schan. Gotha (Petermanns Mitteilungen, Ergänzungsheft 32).

MERZBACHER, Gottfried with collaboration of Alfred LAUBMANN (1907): Wissenschaftliche Ergebnisse der Reise von G. Merzbacher im zentralen und östlichen Tian-Schan 1907/08. München (Abhandlungen der Bayerischen Akademie der Wissenschaften, mathematisch-naturwissenschaftliche Klasse).

MESSERLI, Bruno (1984): Work and History of the Commission on Mountain Geoecology of the International Geographical Union (IGU). In: Lauer, Wilhelm [ed.]: Natural Environment and Man in Tropical Mountain Ecosystems. Proceedings ... Stuttgart (= Erdwissenschaftliche Forschung; 18): 9–11.

MESSERLI, Bruno, Thomas HOFER (1992): Die Umweltkrise im Himalaya. In Geographische Rundschau 44 (7/8): 435–445.

MESSERLI, Bruno, Jack D. IVES [eds] (1997): Mountains of the World. A Global Priority. New York, London.

MESSERLI, Paul (1989): Mensch und Natur im alpinen Lebensraum – Risiken, Chancen, Perspektiven. Zentrale Erkenntnisse aus dem schweizerischen MAB-Programme. Bern.

Mountain Agenda [ed.] (1998): Mountain of the World: Water towers for the 21st century. Bern.

Moyennes montagnes européennes: nouvelles fonctions, nouvelles gestions de l'espace rural. Actes du colloque ... (1999). Clermont-Ferrand (CERAMAC; 11).

PILLEWIZER, Wolfgang (1986): 5 Jahrzehnte kartographischer Arbeit und glaziologischer Forschung. Berlin (Kleine geographischer Schriften; 6).

PRICE, L. W. (1981): Mountains and man: A study of process and environment. Berkeley.

PRICE, Martin (1994): Mountain research in Europe: an overview of MAB research from the Pyrenees to Siberia. Paris (= Man and the biosphere series; 14)

PRICE, Martin (1999): Global change in the mountains. New York.

RATHJENS, Carl (1966): Neuere Entwicklung und Aufgaben einer vergleichenden Geographie der Hochgebirge. In: Geographisches Taschenbuch 1966/69. Wiesbaden: 199–210.

RATHJENS, Carl (1982): Geographie des Hochgebirges. 1. Der Naturraum. Stuttgart (= Teubner Studienbücher Geographie).

RATHJENS, Carl (1988): German geographical research in the high mountains of the world. In: WIRTH, Eugen (ed.): German geographical research overseas. A report to the International Geographical Union. Tübingen.

SCHWEIZER, Günther (1984): Zur Definition und zur Typisierung von Hochgebirgen aus der Sicht der Kulturgeographie. In: GRÖTZBACH, E., G. RINSCHEDE [eds]: Beiträge zur vergleichenden Kulturgeographie der Hochgebirge. Eichstätter Beiträge 12. Eichstätt. 31–55.

SMETHURST, D. (2000): Mountain geography. The Geographical Review 90 (1):35–56.

SOFFER, A. (1982): Mountain Geography – a New Approach. Mountain Research and Development 2 (4):391–398.

STADELBAUER, Jörg (1981): Geoökologische Gebirgsforschung. Ein Bericht über das internationale Symposium der IGU-Kommission "Mountain Geoecology", Japan 1980. In: Erdkunde 35: 321–324.

STADELBAUER, Jörg (1992): Geoökologische Gebirgsforschung als angewandte Umweltforschung. Ein Bericht zur Entwicklung und zu den Zukunftsaufgaben der

IGU-Commission „Mountain Geoecology and Sustainable Development". In: Erdkunde, Archiv für wissenschaftliche Geographie 46: 290–297.
STONE, Peter [ed.] (1992): State of the World's Mountains. A Global Report. London.
TROLL, Carl (1941): Studien zur vergleichenden Geographie der Hochgebirge der Erde. Bonn (= Bonner Mitteilungen; 21). Reprint in: UHLIG, H., W. HAFFNER [eds] (1984): Zur Entwicklung der vergleichenden Geographie der Hochgebirge. Darmstadt (= Wege der Forschung; 223): 128–169.
TROLL, Carl (1959): Die tropischen Gebirge: ihre dreidimensionale klimatische und pflanzengeographische Zonierung. Bonn (Bonner geographische Abhandlungen; 25).
TROLL, Carl (1962): Die dreidimensionale Landschaftsgliederung der Erde. In: Hermann von Wissmann-Festschrift. Tübingen: 54–80; Reprint in Troll 1966: 328–359.
TROLL, Carl (1966): Ökologische Landschaftsforschung und Vergleichende Hochgebirgsforschung. Wiesbaden (= Erdkundliches Wissen; 11).
TROLL, Carl (1966a): Luftbildforschung und landeskundliche Forschung. Wiesbaden (Erdkundliches Wissen; 12).
TROLL, Carl [ed.] (1968): Geo-ecology of the Mountainous Regions of the Tropical Americas. Bonn (Colloquium Geographicum; 9).
TROLL, Carl [ed.] (1972): Geoecology of the high mountain regions of Eurasia. Erdwissenschaftliche Forschung 4. Wiesbaden.
TROLL, Carl (1975): Vergleichende Geographie der Hochgebirge der Erde in landschaftsökologischer Sicht. Eine Entwicklung von dreieinhalb Jahrzehnten Forschungs- und Organisationsarbeit. In: Geographische Rundschau 27 (5): 185–1998.
UHLIG, Harald (1976): Bergbauern und Hirten im Himalaya: Höhenschichten und Staffelsysteme – ein Beitrag zur vergleichenden Kulturgeographie der Hochgebirge. In: 40. Deutscher Geographentag Innsbruck 19. bis 25. Mai 1975. Tagungsbericht und wissenschaftliche Abhandlungen. Wiesbaden 1976: 549–586.
UHLIG, Harald (1984): Die Darstellung von Geo-Ökosystemen in Profilen und Diagrammen als Mittel der vergleichenden Geographie der Hochgebirge. In: GRÖTZBACH, Erwin, Gisbert RINSCHEDE [eds]: Beiträge zur vergleichenden Kulturgeographie der Hochgebirge. Regensburg (Eichstätter Beiträge; 12 Geographie): 93–152.
UHLIG, Harald, Willibald HAFFNER [eds] (1984): Zur Entwicklung der vergleichenden Geographie der Hochgebirge. Darmstadt (= Wege der Forschung; 223).
UNESCO (1974): Working Group on Project 6: Impact of Human Activities on Mountain and Tundra Ecosystems. Final Report, Man and Biosphere Programme, No.14. Paris.
VARJO, Uuno (1985): Subcommission on Rural Development on Highland and High-Latitude Zones: Report for the Period 1980 – 1984. In: LEIDLMAIR, Adolf, Klaus FRANTZ: Environment and Human Life in Highlands and High-Latitude Zones. Proceedings of a Symposium … Innsbruck (= Innsbrucker Geographische Studien; 13): 9–11.
WINIGER, Matthias (1992): Gebirge und Hochgebirge: Forschungsentwicklung und -perspektiven. Geographische Rundschau 44 (7–8):400–407.
WINIGER, M. & J. BENDIX (eds.) (2000): Field Guide for Landscape Ecological Studies in High Mountain Environments: prepared by members of the 'Working Group on Mountain Geoecology', Association of German Geographers (VGDH). Submitted to Commission 96C.16 'Mountain Geoecology and Sudstainable Development' of the International Geographical Union (IGU). Bonn.

Geomorphological Altitudinal Levels of the Mountains and the Influence of the Relief on the Glacier Altitudinal Level during the Quaternary Climate Change as Basis of a Glacier Typology

Matthias Kuhle[1]

Keywords: Relief varieties of the mountains, geomorphological altitudinal levels, glacier altitudinal level, Quaternary climate change, shifting of the snowline (ELA), relief-dependent alteration of glacier types.

Abstract: An overview is given of the essential characteristics of geomorphological altitudinal levels and their varieties in the mountains of the earth. In dependence on the climate, the width and interlocking of altitudinal levels vary very strongly. The characteristics of the glacier altitudinal level, however, the glacier types, are nearly completely dependent on the main characteristics of the mountains, i. e. the relief and its vertical distance.

The fundamental idea is that relief configuration and thus the type of glaciation depend on the location of the snowline (ELA) within a certain landscape. Together with a method of snowline determination the author has developed a calculation scheme with which the characterizing gradient relationship of past and present glaciers can be fixed. The angles α and δ were calculated for 223 glaciers from 12 various mountain regions of the earth and plotted in a scatter diagram.

1 Altitudinal Levels of the Mountains

The global climate zones with their geographic phenomena re-appear in the altitudinal levels of high mountains (Lautensach 1952). So e. g. in Central Europe a meridional shifting of 100 km corresponds to a hypsometric one of 100 altitude metres. This is equivalent to a difference in temperature of 0.5 – 0.6°C. Köppen (1923) has developed a climate system in which for instance both the tundra areas of Greenland and Svalbard at sea level and the dwarf scrub steppes in Tibet about 5000 m a. s. l. belong to the tundra climate. The periglacial-geomorphologic findings fit into this three-dimensional system, too. Forms of patterned ground of arctic dimensions and ice wedges as well as pingos in High Asia between 27° and 39°N (Furrer 1965; An Z. 1980; Zho & Guo 1982; Kuhle 1982, 1985) and between 4200 and 5600 m are proof of a correspondence of subtropical and arctic periglacial phenomena. However, relationship does not mean identity. The effect of solifluction during the time of the day and the time of the year (Troll 1944) with regard to the forms of subtropical mountains and polar latitudes as well as to the different vertical width of this region of forming (Fig. 1) shows differences. Accordingly, comparing the glacier areas close to the sea with those closer to the equator, we have to consider differences caused by the latitude – in spite of the similarity of the metamorphosis of snow into ice and an ice accumulation leading to similar ice streams.

---

[1] Geographie/Hochgebirgsgeomorphologie, Geographisches Institut der Universität Göttingen, D-37077 Göttingen; mkuhle@gwdg.de.

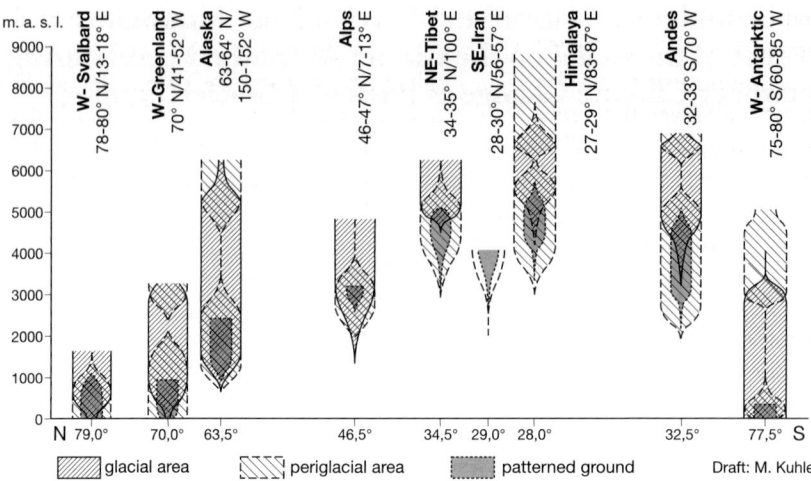

Fig. 1 Nine combinations of glacial and periglacial altitudinal levels in mountains of the earth

The three altitudinal levels of the mountains from bottom to top are the fluvial altitudinal level, the periglacial altitudinal level and the glacial altitudinal level (Fig. 1). They are interlocked so that e. g. fluvial linear erosion is still effective up to a certain height, though on the slopes solifluction already occurs as a characteristic of the periglacial altitudinal level. In the mountains of the earth dynamics of solifluction and patterned ground like these can also be evidenced regularly on slopes and debris faces on both sides of hanging and valley glacier tongues, so that these highest altitudinal levels, too, are interlocked vertically and horizontally.

The Quaternary climate change with the extremes of a glacial-age cooling down by c. 8 – 12°C and an interglacial re-warming, led to a shifting of these climate-geomorphological altitudinal levels in a downward, i. e. upward direction. This shifting becomes especially clear with regard to the altitudinal level of glaciers. Here, a vertical difference of the snowline (ELA = equilibrium line altitude) of 1200 – 1500 m at glacial times compared with today has led to a lowering of the lower limits of this altitudinal level in the mountains of the earth of over 2000 m. Hereby not only the lower limit of the glacier altitudinal level has been lowered, but it has also become wider to this amount, because the higher mountain peaks remained glaciated as during the interglacial period.

2 The Climate-Genetic Variation of the Glacier Altitudinal Level as Expression of the Variety of the Mountain Relief

If due to an Ice Age depression of the temperature about 8 – 10°C (cf. e. g. Kuhle 2005: 193) the snowline (ELA) lowers by c. 1200 – 1500 m in mountains as e. g. the Alps, Alaska Range, Andes, Karakorum, Kuenlun, Tienshan or Himalaya, the nearly twice as strong lowering of the lowest tongue ends of the glaciers makes them much longer and thicker. From this results a possible classification of glaciers which is based on the particular configurations of relief in which the glaciation develops. The fundamental idea in this is that the relief configuration and thus the type of glaciation depend in each case on the location of the snowline (ELA) within a certain mountain landscape. If the snowline (ELA) extends through the summit walls of a mountain range, a summit- or flank glaciation results first. Whether the descending avalanches aggregate and develop perennial firn patches, a wall base or a valley glacier, depends on the location of the wall base. For instance the Doldenhorn N wall in the Alps (Berner Oberland 46°28'06"N/7°44'06"E) extends between 3600 and 1550 m a. s. l. Only the summit region is glaciated and the ELA in this wall lies too high to allow the regeneration of avalanche ice into a glacier. In contrast, the Morgenhorn N wall (46°29'50"N/7°48'24"E) extends between 3600 and 2800 m a. s. l. with the snowline lying at about the same altitude. Here we find a considerable ice regeneration at the wall base building up a debris covered glacier tongue flowing down the valley. An example from the Inner Himalayas is the 6385 m high Pughru Kang (28°50'58"N/83°37'56"E). In spite of the considerable height of this mountain there is merely a flank glaciation and some ice accumulation at the wall base, i. e. the ELA is situated too high up in the wall for the formation of a regenerated glacier: in contrast, the Sabche Kang (Annapurna III E summit 28°33'41"N/84°03'42"E Himalaya) is 6742 m high and in vertical direction has 550 m more catchment area. Thus, a considerable glacier collects at the wall base and flows down into the valley. The author named this type "avalanche cone glacier" (KUHLE 1982) (see Fig. 6 I.a) and considers it to be the most characteristic recent glacier type in the Himalayas. In case of an ELA lying nearer to the wall base the next genetic step is reached by the formation of an "avalanche caldron glacier" (terminology according to SCHNEIDER 1962 (Fig 6 I.b). The author's example for this type is the 8 km long Horcones Inferior glacier, which is nourished solely by avalanche cones from the highest wall of the American continent, the Aconcagua south face (Photo 1). As indicated by the debris cover, the entire glacier surface is situated below the snowline. If the snowline happens to descend even below the valley heads of smaller tributary valleys we are faced with a glacier type which, according to VISSER (1934) and SCHNEIDER (ibid.), must be called a "firn caldron glacier". Examples are the Unterer Grindelwaldgletscher (Alps 46°33'56"N/8°06'07"E), the lowest left tributary glacier (Photo 2) of the K2 N glacier in the Karakorum N slope and the Nun-Kun N glacier (34°00'49"N/76°00'19"E) in the Zanskar Himal. A further lowering of the ELA – making more of the relief's vertical distance accessible for glaciation – leads to a type called "firn stream glacier" (see Fig. 6 II.b). Examples are the S-Innelschek glacier (Photo 3), the Fedtschenko glacier (38°36'42"N/72°18'30"E W Pamir), the Cerro Juncal SW glacier (33°07'07"S/70°04'03"W Andes, S America) or the Upper Muldrow glacier (Mt. McKinley 63°05'09"N/150°47'14"W, Alaska). Here, parts of the actual valley glacier are located above the snowline and thus form a part of the catchment area. According to SCHNEIDER (1962) the "ice stream system" is defined by glaciers having a common catchment area separating and flowing down in different directions and/or valleys. Thereby this type only represents a quantitative and not a qualitative difference to the "firn stream glacier". The "ice stream system" with "firn stream glaciers" also act, in part, as a transition to inland glaciations, e. g. in W Spitsbergen (Photo 4). The following glacier type, the "firn basin glacier" is characteristic of the Alps (Fig. 2), but also occurs e. g. in the Andes (Plomo glacier 32°58'29"S/70°00'38"W Tupungato group), Inner Himalayas, Tibet (Fig. 6 II.b) or W Greenland. In this system of decreasing relative differences in height, that is an increasing surface area at a decreasing vertical distance, and the corresponding filling of the mountain – i. e. valley relief, this glacier type follows the "firn stream glacier". The surface of the "firn stream glacier" is still very clearly set into its valley receptacle, whilst the surface of the "firn basin glacier" is merely flatly inset between the intermediate valley ridges. Its ice has nearly filled the valley receptacle. The typologically last step in this system is the "central firn cap". This type can be found not only in polar regions, but also in high mountains, so e. g. at 5200 m a. s. l. at 33°02'56"S/69°58'15"W in the Central Andes (plateau glacier near Plomo glacier), in NE Tibet (Dunde glacier 38°05'43"N/96°26'02"E Kakitu massif) or between 3000 and 3200 m a. s. l. in the European Alps (Marmolata glacier 46°26'06"N/11°51'49"E). The type "central firn cap" also includes the "ice sheet" or "inland glaciation" which covers the landscape completely (NE-Land of Spitsbergen, Greenland, Antarctica). Here the ice surface extends over vast horizontal distances with very small vertical distances.

The basic idea of this classification can be illustrated best by the example of Dhaulagiri I (28°41'45"N/83°29'43"E). Its 4600 m high W face is sloping from 8172 to 3550 m a. s. l. This wall rises c. 3000 m above the snowline but in spite of this large catchment area only a very small glacier is regenerated at the wall base, because this region lies too far below the snowline. On the other hand, at the south face of Dhaulagiri the wall is just 4000 m high which means the base is c. 600 m higher than on the west face. Consequently, the avalanche ice is regenerated at the wall base into an extensive glacier terrace, even having a glacier tongue. The N face of Dhaulagiri I is only 3500 m high, i. e. compared with the S face is again 500 m higher and even 1100 m higher in relation to the W face. This results in a valley glacier 17 km in length, in spite of the fact that at the north face the orographic snowline is located 300 m higher than at the south face owing to the lack of monsoon precipitation. The summits in the Dhaulagiri and Annapurna massifs reach up to and over 8000 m, and the valley floors are very deep, some of them below 3000 m a. s. l. Thus vertical distances between 4000 and 5000 m or even more do occur. Formerly, during the last glacial period (LGP and LGM), the valleys were filled only with a relatively small amount of ice – merely the types of "firn caldron glacier" and "avalanche caldron glacier" were realized (see KUHLE 1982). In contrast, the vertical distances in the area of the Inner Himalayas are much smaller. Here we have summits at

Photo 1   S-wall of Aconcagua

Photo 2   View from the orographic right-hand valley flank across the K2 parent glacier

Photo 3   Second valley glacier of the S Innelschek glacier

about 6000 m, and the valley floors are barely below 4000 m, i. e. the vertical distance amounts to 2000 – 2500 m at the most. Consequently, this region was covered by an ice stream system during the last glacial period (LGP, LGM) (KUHLE 1982).

Still further to the N, in the Tibetan area, the vertical distance between the summits of the mountains set upon the high plateau and the height of the valley bottom and the plateau decreases further. So for instance in the Central-Tibetan Tanggula Shan the summits are 6000 – 6500 m high (Mt. Geladaindong 33°30'01"N/91°04'E), whilst the valley floors and mountain forelands are situated about 5400 m (KUHLE 1991). Accordingly, here an "ice stream system" at an ELA about 5700 m a. s. l. exists still today (cf. Fig. 6 II.b). During the last glacial period (LGP, LGM) an "ice sheet" or "inland glaciation" covered the mountain landscape nearly completely (KUHLE 2004).

Photo 4   James I and Oscar II Land on Spitsbergen

The main chain of the High Himalayas now as well as in former times never experienced anything but a "firn caldron glaciation" and "avalanche caldron glaciation", although the ice was much thicker and the ice cover of a greater extent. However, in the Inner Himalayas in spite of much smaller vertical distances in the catchment areas (Fig. 6), the water divides were superposed by an "ice stream system" and thus showed a typologically more massive glaciation than the High Himalayas (e. g. Fig. 6 II.b). In Tibet this Ice Age glacier type of the Inner Himalayas even develops today (see above), because the ELA in relation to the relief has still further been lowered (Fig. 6 II.b). Presently it runs only just 300 m above the basal face, i. e. the height of the Tibetan plateau. During the Ice Age the most extended glacier type at all – an "inland glaciation" – has existed there, because the ELA has remained even below the height of the plateau.

3   A Glacier Typology in Dependence on Specific Relief Parameters

Together with a method of ELA determination the author has developed a calculation scheme with which the characterizing gradient relationships of former and present glaciers can be fixed (KUHLE 1988). The procedure will be demonstrated using the example of the Aletsch glacier (Fig. 2, 3). Here the average value of all tributary culminations (only summit points) over the base value (= mathematical average between the highest summit, the Jungfrau, 4158 m a. s. l., and the glacier terminus in the year 1960, 1503 m a. s. l.) is considered to be the level of the glacial catchment area. Imitating the method of v. HÖFER (1879), the mathematical snowline (s(m)=2689 m) is yielded from the mathematical average between the average catchment area altitude (3875 m) and the glacier terminus (1503 m). From the height s(m) the horizontal distance (b) is measured perpendicular to the direction of flow of the glacier towards the highest point of culmination, but only up to the average catchment area level; in case of a compound glacier this must be done for every major tributary. The horizontal (e) from the glacier terminus to s(m) is calculated in the same way. The inclination angles α and δ can be calculated from the respective vertical and horizontal distances by means of the tangential function (Fig. 3). The angles α and δ were calculated for 223 glaciers from 12 various mountain regions of the earth and plotted in a scatter diagram (Fig. 4 α = y-axis; δ = x-axis).

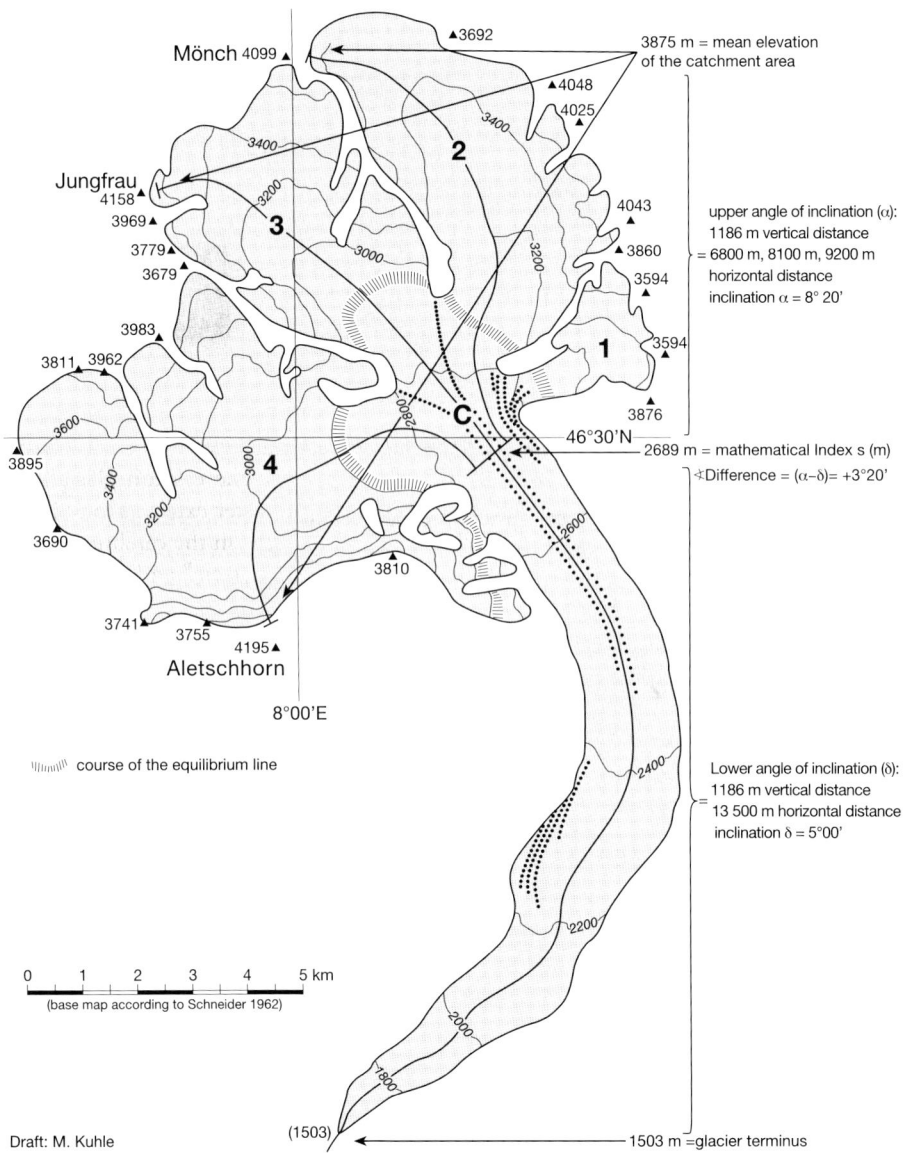

Fig. 2   Determination of the angles α and δ using the Aletsch glacier

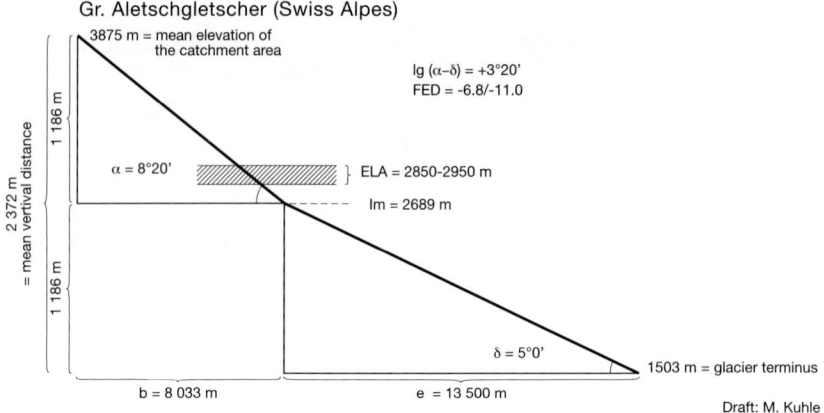

Fig. 3 The angle and snowline (ELA) relationships of the Aletsch glacier in cross section

To make the numerical classification phenomenologically, i. e. conceptually understandable, the glaciers were also labelled with the terminology of Schneider (1962). The distribution shows that a meaningful glacier typology is produced with the mathematical classification (α : δ). The idealized scheme is presented in Figure 5. It becomes clear that the individual phenomenological types form clearly defined groups which, however, also overlap on the peripheries.

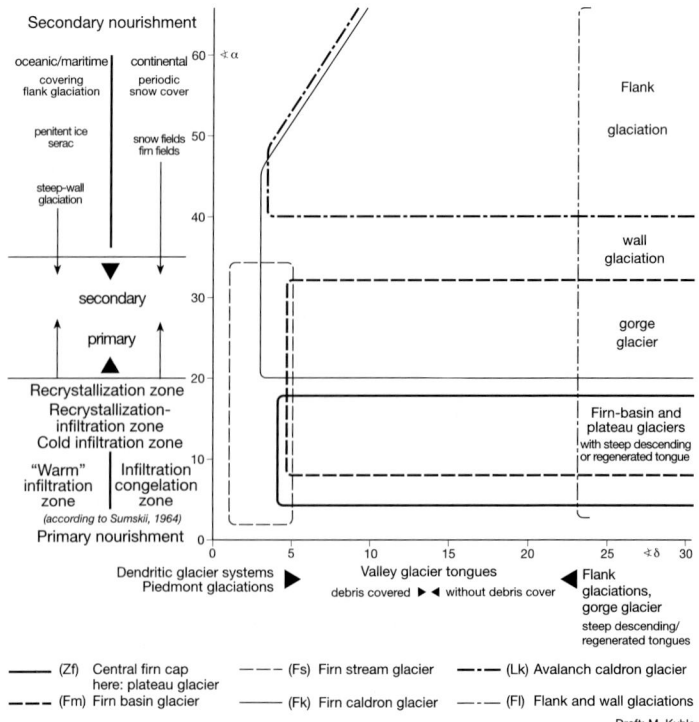

Fig. 4 Glacier-typological scatter diagram from 223 modern glaciers

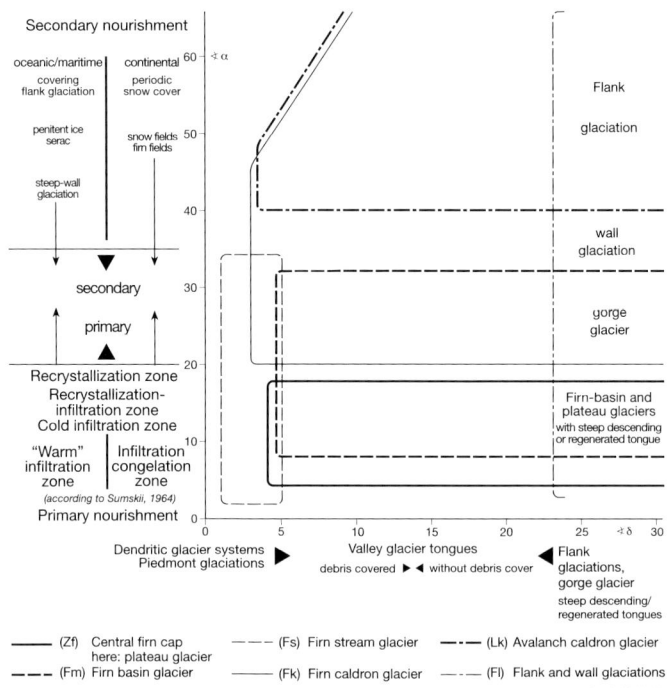

Fig. 5 Glacier typology scheme based on the angles α and δ

By doing so, the aforementioned empirical type continuity and combinations, which make it difficult to classify a glacier in homogeneous clearly defined groups, is presented in the correct way. By means of the α value, an assessment of the mode of glacial nourishment is possible: up to $\alpha = 20°$, primary glacial nourishment is dominant; between

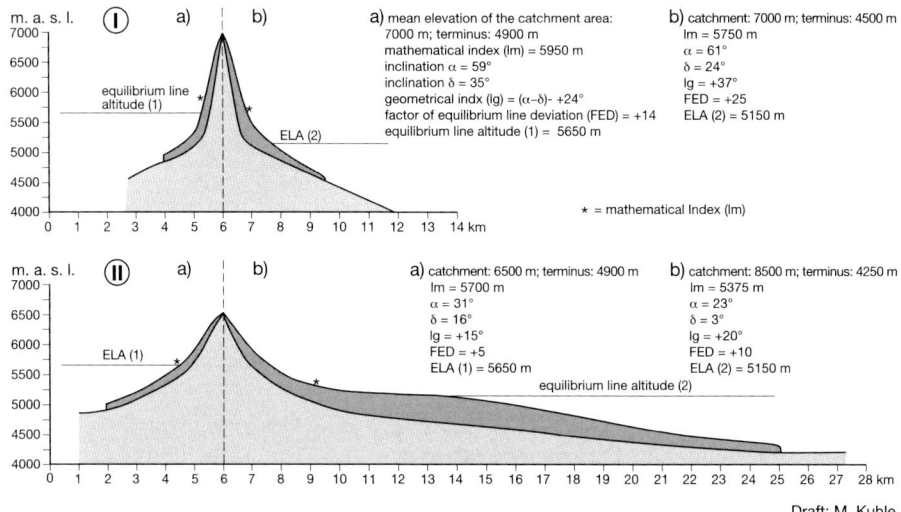

Fig. 6 Dependency of the intensity of glaciations on topography

20° and 35° the portion of secondary nourishment increases, which dominates above 35° (Fig. 4, 5). The thermodynamic classification of ice formation processes according to Shumskii (1964) is also specified here to illustrate that the catchment areas can phenotypically show completely different ice covers even with an identical angle α. For example, when a high continental mountain glacier with an angle α = 40° attains the recrystallization zone with its catchment area, periodic snow cover and snow fields can be found here (Kuhle 1986, 1988); however, when a glacier with angle α = 40° is situated at a low altitude in maritime mountains, covering flank glaciations and steep-wall glaciers do occur. The values of δ give indications as to the configuration of the glacier tongues (Fig. 5). The numerical classification of glaciers using the angles α and δ has the following characteristics: (a) the localization of individual glaciers within the scheme is clearly determinable and free of subjective decisions; (b) the diagrammatic assignment of glaciers does not signify a necessary assignment to a homogeneous type, but enables specifying statements on the position within the phenomenological convergence zones; (c) the mathematical classification is applicable to present and past glaciations with equal precision and thus enables an absolute comparability.

References

An Zhongyuan (1980): Formation and evolution of permanent ice mound on Quing-Zang Plateau. In: J. Glaciology and Cryopedology 2, No. 2, Lanzhou: pp. 25–30.

Furrer, G. (1965): Die Höhenlage von subnivalen Bodenformen. Habil., Pfäffikon: pp. 1–78.

Höfer, H. v. (1879): Gletscher- und Eiszeitstudien. Sitzungsbericht der Akademie der Wissenschaften Wien. Mathematisch-physikalische Klasse I 79: 331–367.

Köppen, W. (1923): Die Klimate der Erde. Berlin.

Kuhle, M. (1982): Der Dhaulagiri- und Annapurna-Himalaya. Ein Beitrag zur Geomorphologie extremer Hochgebirge. Z. f. Geomorph. Suppl. Bd. 41, Bd. I (Text): 1–229, Bd. II (Abb.): 1–183 u. Geomorph. Karte 1:85 000.

Kuhle, M. (1985): Permafrost and Periglacial Indicators on the Tibetan Plateau from the Himalaya Mountains in the south to the Quilian Shan in the north (28-40°N). In: Z.f. Geomorph. N. F., 29, 2: pp. 183–192.

Kuhle, M. (1986): Die Obergrenze der Gletscherhöhenstufe – Oberflächentemperaturen und Vergletscherung der Himalayaflanken von 5000 – 8800m. Z. f. Gletscherkd. u. Glazialgeologie 22 (2): 149–162.

Kuhle, M. (1988): Topography as a Fundamental Element of Glacial Systems. A New Approach to ELA-Calculation and Typological Classification of Paleo- and Recent Glaciation. GeoJournal 17, 4, Tibet and High-Asia, Results of the Sino-German Joint Expeditions (I): 545–568.

Kuhle, M. (1991): Observations Supporting the Pleistocene Inland Glaciation of High Asia. GeoJournal 25, 2–3, Tibet and High Asia, Results of the Sino-German Joint Expeditons (II): 133–233.

Kuhle, M. (2004): The High Glacial (Last Ice Age and LGM) ice cover in High and Central Asia. Development in Quaternary Science 2c (Quaternary Glaciation – Extent and Chronology, Part III: South America, Asia, Africa, Australia, Antarctica. Eds: Ehlers, J.; Gibbard, P.L.): 175–199. (Elsevier B.V., Amsterdam).

Kuhle, M. (2005): The maximum Ice Age(Würmian, Last Ice Age, LGM) glaciation of the Himalaya- a glaciogeomorphological investigation of glacier trim-lines, ice thicknesses and lowest former ice margin positions in the Mt. Everest-Makalu-Cho Oyu massifs (Khumbu and Khumbakarna Himal) including informations on late-glacial, neoglacial and historical glacier stages, their snow-line depressions and ages. GeoJournal 62, 3–4. Tibet and High Asia (VII): 191–650.

Lautensach, H. (1952): Der geographische Formenwandel. Studien zur Landschaftssystematik. Colloquium Geographicum 3 Bonn.

Schneider, H. J. (1962): Die Gletschertypen. Versuch im Sinne einer einheitlichen Terminologie. Geographisches Taschenbuch 1962/63: 276–283.

Shumskii, P. A. (1964): Principles of Structural Glaciology. Translated from the russian by D. Kraus. Dover Publ. Inc., New York. (Translated from russian by D. Kraus).

Troll, C. (1944): Strukturböden, Solifluktion und Frostklimate der Erde. Geol. Rundschau, 34: pp. 545-694.

Visser, P. C. (1934): Benennung der Vergletscherungstypen. Z. f. Gletscherkd 21: 137–139.

Zho Youwu & Guo Dongyin (1982): Principal Characteristics of Permafrost in China. In: J. Glaciology and Cryopedology 4, 1. Lanzhou: pp. 1–19.

Figures and Photos

Fig. 1: Nine combinations of glacial and periglacial altitudinal levels in mountains of the earth.

Fig. 2: Determination of the angles α and δ using the Aletsch glacier (46°30'07"N/8°01'58"E, Alps, Berner Oberland). The main catchment area is formed by Ewigschneefeld (2), Jungfrau Firn (3) and Großer Aletsch Firn (4), whereas Grunegg Firn (1) is of subordinate importance for nourishment and does not enter into the calculation of the angles α; the same applies for the Mittlere Aletsch Gletscher; C is the Concordia Platz.

Fig. 3: The angle and snowline (ELA) relationships of the Aletsch glacier in cross section (see Fig. 2.)

Fig. 4: Glacier-typological scatter diagram from 223 modern glaciers (after Kuhle 1988) based on the angles α and δ (cf. Fig. 2, 3).

Fig. 5: Glacier typology scheme based on the angles α and δ.

Fig. 6: Dependency of the intensity of glaciations on topography: No.I. corresponds to the relief of the main chain of the Himalayas, No.II. to the mountain relief within the central Tibetan plateau (e. g. Tanggula Shan). Under equal climatic conditions the glacier termini of I.a) and II.a) reach the same altitude although the feeding ground of I.a) is 500 m higher; this can be put down to the smaller accumulation capacity of the steeper feeding ground of I.a) (α = 59°). When reducing the ELA by 500 m I.b) only reaches 4500 m while II.b) flows down to 4250 m (further details see Kuhle 1988).

Photo 1: Taken at 4000 m a. s. l. from the left side of the Quebrada Horcones Inferior seen to the NNE into the S-wall of Aconcagua (6996 or 7021 m; Mendoza Andes). The wall drops for some 2800 metres. The ridge does not descend below the 6800 m level at any point. (□) is the 1.8 km broad upper Horcones Inferior glacier (32°40'44"S/69°58'45"W); it is an "avalanche caldron glacier". The ice avalanche cones (V) at its base (a little hidden by the angle of view) terminate at 4200 m ca. 600 m below the ELA. At times, over the steps, outcrops of rock break through the ice-fields. Ice avalanches break down from ice balconies formed on the edges of ice fields in the wall (△) and fall down to the wall foot (V). There they regenerate into glacier ice and nourish the valley glacier. (⇧) marks the glacier ice. Its surface is completely covered by melted surface moraine (□). (■) is its orographic left lateral moraine. Analogue photo M. Kuhle.

Photo 2: Viewpoint 5300 m a. s. l.; Karakorum-N-slope. View from the orographic right-hand valley flank across the K2 parent glacier (■ white) towards the WNW to the left side tributary glacier (■ black; 35°57'50"N/76°26'58"E) furthest down the valley. This glacier (■ black) is a "firn caldron glacier" because the two feeding areas of the glacier are two caldrons (●) situated 250 – 300 m above the ELA. Analogue photo M. Kuhle.

Photo 3: At 5200 m a. s. l. facing S looking across the second valley glacier of the S Innelschek glacier connected from the E on the orographic left side (□ 42°05'13"N/80°08'27"E Tienshan). This is a source branch of the S Innelschek glacier. This glacier and several western and eastern parallel glaciers show the classical characteristics of the type "firn stream glacier". The surface of this valley glacier (□) is situated above the ELA, so that it is nourished mainly by primary snow precipitation. Secondary nourishment is provided by avalanches pouring down into the connected hanging valley (O). The summit which makes up the orographic left flank of the glacier valley, is a 6820 m high E satellite of Pik Pobedi. Analogue photo M. Kuhle.

Photo 4: Taken at 590 m a. s. l. from the Kongressfjellet (Dickson Land) looking WNW on to James I and Oscar II Land on Spitsbergen. The centre of this arctic mountain glacier system (Mt. Centralen, on the right beside the right ✘) is situated at 78°42'07"N/13°25'05"E. Typologically it has to be classified as an "ice stream system", because the levels of the large valley glaciers of this peninsula and their nourishing areas are connected with each other across the ice divides without any breaks in slope worth mentioning. On this side and also on the other side of the mountains the valley glacier tongues calve into the sea (⇧). The connected valley glaciers with their small side valley- and cirque glaciers (✘) fill the mountain relief with their thick ice on a large scale. Analogue photo M. Kuhle.

Diversity is in the Eye of the Beholder – Plant Diversity Patterns and their Relation to Scale: Examples from the South-Western USA

Friederike Grüninger & Thomas Fickert[1]

Geographic patterns of plant diversity and the underlying reasons why some areas contain higher species numbers than others, have attracted ecologists since early accounts by Alexander v. Humboldt and Charles Darwin. Them and their descendants originally treated the term diversity simply as a synonym for species richness (McIntosh 1967, Peet 1974) thereby reflecting the information best available rather than endorsing the term's adequacy in regard to the complexity of the matter.

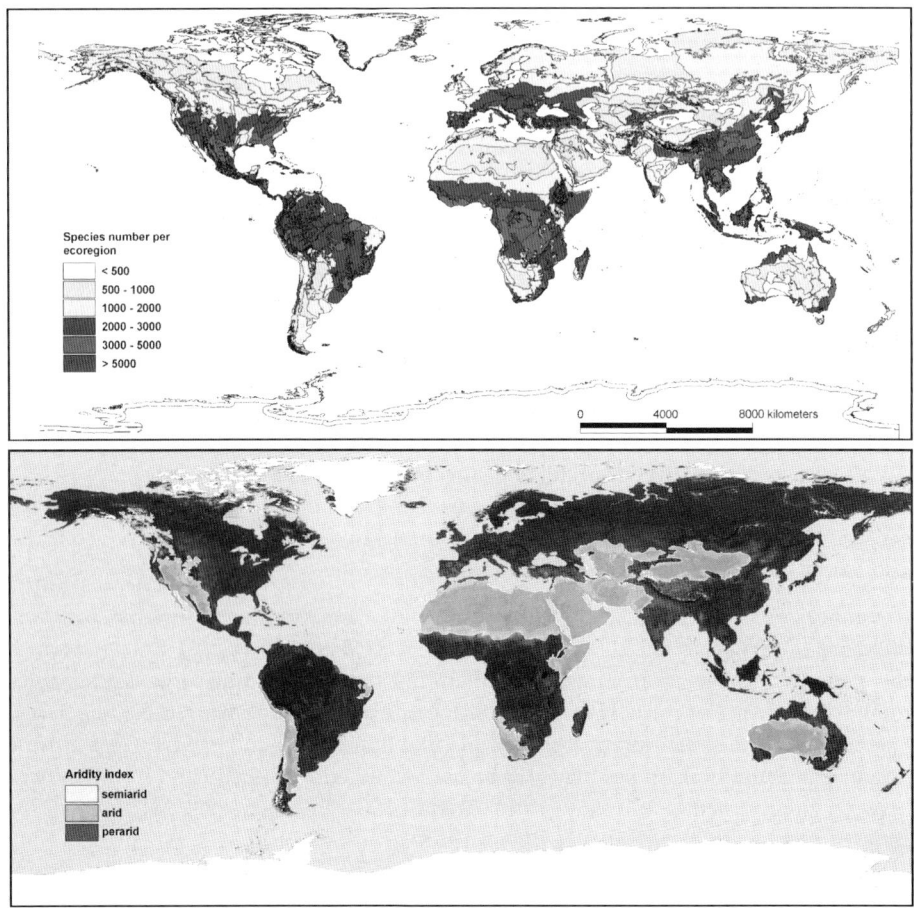

Sources (a): http://www.uni-bonn.de/Aktuelles/Presseinformationen/2005/166/bilder/figure1a.jpg; (b): http://www.unep.org/geo/news%5Fcentre/gallery.asp

Fig. 1   Global patterns of plant diversity per ecorogion (a) and the distribution of dry regions (b)

[1]   University of Passau, Physical Geography, D-94032 Passau; friederike.grueninger@uni-passau.de; thomas.fickert@uni-passau.de.

Still, mapping complex dimensions of biological diversity is complicated and rarely done (WILSON 1985, GASTON 1998), but during the last decade, a rising number of data sets or improved data has successively led to more detailed maps of global plant species richness (BARTHLOTT et al. 1996, BARTHLOTT et al. 1999, BARTHLOTT et al. 2005, BARTHLOTT et al. 2007, KREFT & JETZ 2007). From a global perspective, an increase in species number between poles and the equator is apparent in these maps. This trend is most obvious for the American Continents (e. g. MUTKE & BARTHLOTT 2005, BARTHLOTT et al. 2007), but in other areas is obscured and modified by either drylands (negatively) or highlands (positively). Mountain ranges in all parts of the world possess higher species number than the surrounding lowlands, whereas increasing aridity generally causes a decrease in niche diversity and thus species number. Interestingly, of all the major desert areas of the world the dry regions within the North American Southwest are surprisingly rich in vascular plant species (Fig. 1, see also BARTHLOTT et al. 2007). The ecological reasons behind this plant diversity and geographical patterns involved shall be considered in this paper. In doing so, we treat the Southwest in a somewhat broad-minded fashion to include all the area between the 42°N (i. e. excluding Oregon and Idaho) and the Mexican Border and between the two main Cordilleras of the North American continent.

2   Diversity – Where to Look and What to Expect

Focusing on vegetation as the primary component of terrestrial biodiversity, plant diversity clearly incorporates more quantitative and qualitative parameters than just the number of taxa in a given environmental unit. The 1992 UN Convention on Biological Diversity in Rio de Janeiro understood biological diversity as "the variability among living organisms from all sources, including all terrestrial, marine, and other aquatic ecosystems and the ecological complexes of which they are a part; this includes diversity within species, between species and of ecosystems" (*UN* 1993, Article 2, p. 146). Barthlott et al. (1996) incorporate furthermore the sum of genetic diversity among organisms, their abundance and their equitability within a specific area, and thus consider also spatial and structural diversity dimensions.

If components from the field of vegetation ecology are considered, biomass and productivity (WHITTAKER 1970, 1972), canopy cover (BARBOUR et al. 1999), as well as the richness of vegetation strata and functional types (life forms, leaf types, structural parameters, etc.; see HUSTON 1994) can also be subsumed under the term. Furthermore, other quantitative parameters like dominance and evenness patterns of taxa or plant functional types may also be used to describe diversity and its differences between environmental units and scales (AUSTIN 1999, BARTHLOTT & WINIGER 1998, BEIERKUHNLEIN 1999, WILSON & PETER 1988).

Obviously, addressing diversity issues for a given area usually does not result in solid short answers. Excellent and encompassing approaches to describe, and attempts to quantify biological diversity, were presented by HUBBELL (2001), HUSTON (1994), RICKLEFFS & SCHLUTER (1993), ROSENZWEIG (1995) and WILSON (1985) among others. The approaches vary, yet, most authors agree that there will be no simple, universal explanation for the spatiotemporal patterns and processes observed. As HUSTON

(1994, p. 7) states, "the challenge is to identify the conditions (e. g. spatial and temporal scales, evolutionary and geological history, disturbance regimes, resource availability, etc.) under which specific mechanisms are likely to have the greatest influence on the diversity of specific functionally-based subsets of organisms."

2.1 The Establishment of Order – Whittaker's Diversity Concept

The classification of different diversity types and different observation scales is an important tool to establish order within this diverse research field. By the differentiation of alpha-, beta-, and gamma-diversity, WHITTAKER (1972, 1977) interlinked different types of diversity and their dependence on spatial scales and environmental gradients.

Species richness per area unit is treated as alpha-diversity. The spatial extension of the sampled area may vary, but the plots are always defined as homogeneous entities within the units under consideration – i. e. spatial, temporal or functional units). The diversity of a larger area composed of several of those basic units is defined as gamma-diversity. These two diversity types may thus have the same dimensional characters with discrete numbers and the same way of calculation. Therefore, all spatial dimensions are covered by these measures of "within-habitat diversity" – WHITTAKER 1972).

In contrast, the beta-diversity or "between-habitat diversity" – WHITTAKER 1972) is based on the previously ascertained quantities and gives comparative information about the similarity –or dissimilarity) between two units; Again spatial, temporal or functional turnovers are measurable – MACARTHUR 1965). Beta-diversity has no own dimension and is described by indices like the "Jaccard"- or "Sørensen-index", or by the "beta-turnover" – WILSON & SHMIDA 1984).

2.2 The Reasons Behind – Driving Forces for Plant Diversity

The plant diversity of a particular region is controlled by a group of different factors, which are commonly subdivided into (1) environmental factors, (2) recurrent dynamics in vegetation structure (sensu WHITE 1979), and (3) time (PAUSAS & AUSTIN 2001).

The environmental factors (1) themselves include resource variables (all properties that are consumed by plants such as light or water, with plant growth being generally greatest at high resource levels), direct variables (e. g. air temperature or soil pH which regulate the physiological processes of the plants but are not consumed, with plant growth being highest at intermediate levels) and indirect or "complex" variables (e. g. altitude, topography, which do not influence plant growth directly, but are correlated with a variety of resource and/or regulator variables; for details see HUSTON 1994).

Recurrent dynamics (2) are subdivided into exogenous disturbances (natural and anthropogenic), stress, and endogenous cyclic processes. Common natural disturbances in the Southwest include wildfire, mass movement, frost-induced soil movement, animal impacts and gap dynamics. Human activity is generally low above the foothills, but the partial use of natural resources (timber, recreation, water usage) and man made fires have some influence. Stress includes processes, which occur regularly and caused the vegetation to adapt (e. g. avalanche gulches, ravines or active alluvial fans). Endogenous cyclic processes are caused by the system itself and alter the vegetation periodically.

Overlying all other processes is the factor of time (3): the temporal variation of plant diversity is controlled by episodic, periodic or permanent changes of direct and/or regulator variables or biotic factors. In addition, the floristic history of a particular region (i. e. the time available for speciation or the occurrence of events of extinction) plays a major role for the species pool available.

2.3 Scale Matters!

Scale issues as scientific topic in relation to diversity research came to the forefront within the last 20 years (TURNER et al. 2001). Depending on the scale of observation

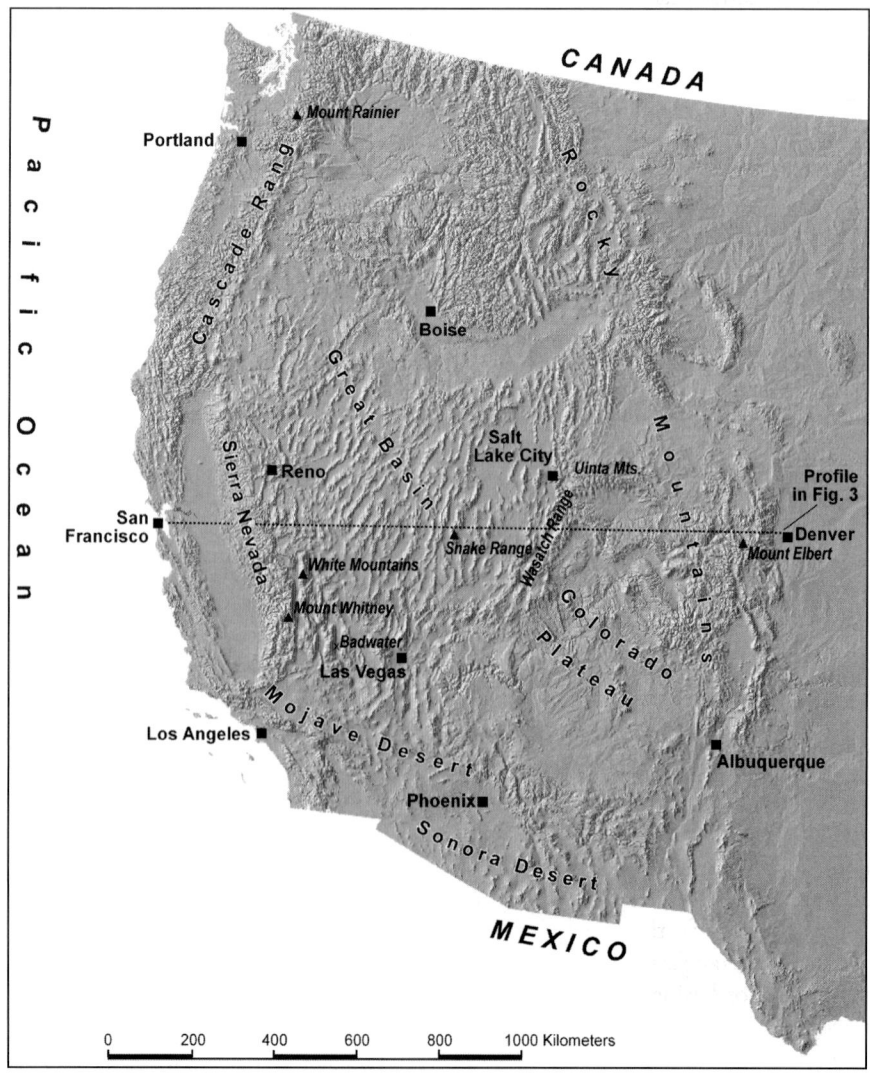

Fig. 2   Shaded relief model of the Western Half of the United States of America (modified from THELIN & PIKE 1991); included are important localities mentioned in the text (modified from GRÜNINGER & FICKERT 2003)

(generally reflecting extent as well as grain sensu TURNER et al. 2001), diversity–related patterns or processes might become prominent or hidden. LEVIN (1992, p. 1960) states, "that there is no single correct scale or level at which to describe a system does not mean that all scales serve equally well or that there are not scaling laws". Factors with a certain spatial or temporal extent create different sized habitat types with different diversity patterns. Mechanisms that affect diversity on one scale might not be effective or visible on another, and a functional interpretation always has to be orientated on the different spheres of influence (CRAWLEY & HARRAL 2001, HUSTON 1994).

Therefore, the scale of observation is crucial for the determination of the relative importance of certain factors (KENT 2005, WILLIS & WHITTAKER 2005). For instance, dynamic processes such as forest fires, avalanches or mass movements strongly enrich the niche diversity and as such are important drivers to increase diversity on finer scales (e. g. considering particular altitudinal vegetation belts with finer grain size). However, as such disturbances are widespread and the taxa occupying the created niches are present in the area anyway, they will barely raise the gamma-diversity of larger areas. On such coarser scales of examination, diversity is mainly controlled by "geodiversity" (sensu BARTHLOTT et al. 1996) as the sum of abiotic factors such as climate (and its changing parameters along vertical, latitudinal and/or meridional gradients), vertical extension, petrography, soils, etc. The following paragraph intends to show examples from the South-Western USA for this scale dependant conspicuousness of drivers for plant diversity.

3 Applying the Concepts – Examples from the South-Western USA

3.1 Setting the Stage – an Outline of the Natural Environment of the South-Western USA

The landscape of the western part of the United States is dominated by two major mountain systems, the Sierra Nevada-Cascades-axis in the far West and the Rocky Mountain System several hundred kilometers to the east (Fig. 2). Although traces of earlier orogenies exist, the present day topography in the western part of the US is caused exclusively by tectonic process going on since late Tertiary times. While subduction processes cause active volcanism in the Cascade Range, the Sierra Nevada, large parts of the Rocky Mountains as well as the vast region in between are a result of block tectonic and uplifting since Miocene times.

This huge region between the two main cordilleras is commonly addressed as Intermountain Region. However, it is not only lying between mountains but consists of mountainous terrain itself with many mountain tops culminating well over 3500 m a. s. l. Its western half – the Great Basin – is characterized by countless North-South trending mountain ranges separated by valleys and basins in between (Fig. 2). This strip-like arrangement of basins and ranges is a result of fault block tectonic which occurred parallel to the Sierran uplift since late Tertiary. Extensional processes led to a cracking of the crust, and the displacement of large blocks created the rising north-northeast trending mountains (STOKES 1986). As a consequence of these tectonic processes, a wide variety of sedimentary, igneous and metamorphic rocks of different ages (up to Paleozoic) are exposed at the surface today.

The Colorado Plateau in the eastern half of the Intermountain Region is quite different and (with the Wasatch Range and the Uinta Mountains) somehow connected to the Rocky Mountains (Fig. 2). Instead of the horst and graben structures present in the Great Basin, here, larger plateaus were uplifted to various levels of altitude. Colorful

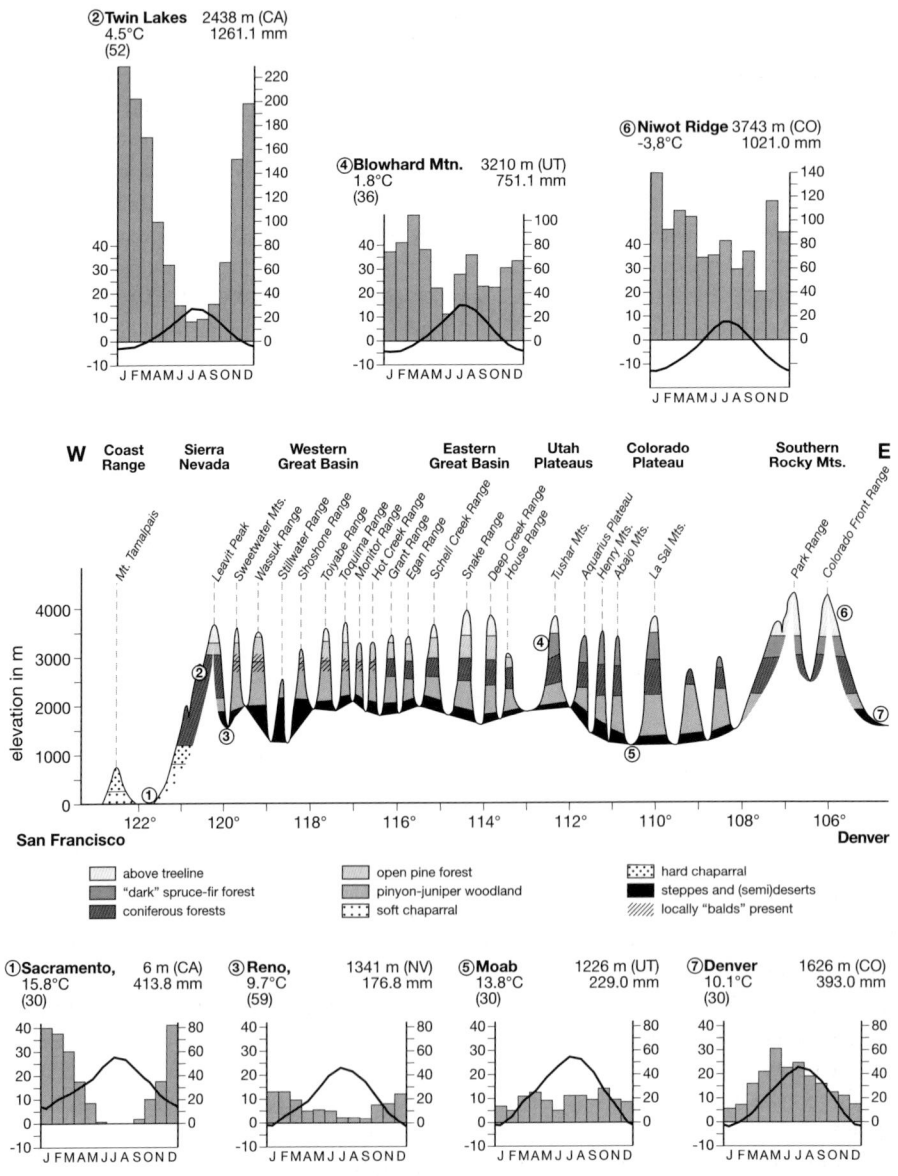

Fig. 3 Schematic diagram showing general types of arrangement of elevational belts in mountain systems (modified from RICHTER 2000).

sedimentary layers from the Mesozoic to Cenozoic era are deeply incised by streams, most impressively in the famous Grand Canyon. Coincidental with the extension and uplift in the Intermountain Region, volcanic activity arose locally.

Climatically, the western part of North America is controlled by the seasonal shift of two dynamic pressure systems – the Aleutian Low and the subtropical eastern Pacific High. In addition, two seasonal thermal pressure systems over the continent – high pressure in winter and near-surface low pressure in summer over the south-western drylands – strongly influence and modify the direction of the western air streams. According to Weischet (1996), three areas can be distinguished: south of 35°N, the Pacific anticyclone prevails almost year-round, causing low precipitation and high temperatures – only during winter the high pressure is slightly weakened permitting cyclonic fronts to enter; north of 45°N, western air streams provide precipitation year-round with only a short decline in summer; between those two areas, the seasonal alteration in the position of the Pacific anticyclone and the Aleutian low pressure creates a Mediterranean-type climate with mild, humid winters and dry, hot summers (Fig. 3).

Besides the circulation patterns, topographic effects are important for the climatic conditions in the Southwest. The prevailing westerlies are blocked by the Sierra Nevada–Cascades-axis and forced to drop most of the moisture on the windward side (Bryson & Hare 1974). On the leeward side, foehn effects cause low vapor content, decreased precipitation and high incoming and outgoing radiation. Thus, semi-arid to arid, highly continental conditions exist to the east of the Pacific Cordillera with mean annual precipitation values in the lowlands between 50 and 300 mm (Weischet 1996). Even at higher elevations the annual precipitation remains below 500 mm in the immediate rain shadow of the Sierra Nevada, (e. g. Station Barcroft above 3800 m a. s. l. in the White Mountains, Powell & Klieforth 1991).

Apart from the decline in amount, the annual course of precipitation in the western part of the Intermountain Region still resembles a Mediterranean-type climate with maximum values in winter and dry summers. However, to the east and southeast the precipitation peak successively shifts into the summer months (Fig. 3). This is caused by the increasing influence of tropical humid air masses originating from trade winds crossing the Gulf of Mexico and from 'gulf surges' from the Gulf of California (Bryson & Hare 1974, Sheppard et al. 2002). Due to thermal low pressure above the high plateaus and mountain bodies of the Intermountain Southwest, warm and humid air is diverged inland and rises above the heated surface of the mountain masses, causing short but intense convective thunderstorms ("Merriam effect", see Lowe 1961). Besides the strong latitudinal and longitudinal climatic variation, high-relief accounts for pronounced altitudinal climate gradients. While temperature lapse rates do not vary significantly in the area, there are strong differences in the altitudinal change of precipitation and thus in the number of arid months (Fickert 2006) due to location and exposure.

The wide spectrum of latitudinally, longitudinally and altitudinally triggered climatic conditions is reflected by the varied vegetation, ranging from steppe and semi-desert types in the dry lowlands to dense coniferous forests and alpine mats at more humid higher elevations (Fig. 3). Due to the pronounced rain shadow effect, the Sierra Nevada shows a strong asymmetry of vegetation belts between western and eastern slopes, especially at lower and middle elevations. Shrub steppes and Pinyon-woodlands on the eastern sides are contrasted by chaparral communities and/or dense conifer forests on the windward side of the Sierra Nevada. Only at higher elevations clouds lap over the crest and weaken the differences in moisture availability, thereby causing high taxonomic similarities between both expositions from the upper conifer forest belts upward.

Within most of the Intermountain Region, the basins and lower foothills are covered by shrub steppes dominated by several *Artemisia* spp. In the lower and warmer southern deserts, *Larrea tridentata* or *Coleogyne ramosissima* are important elements, in parts associated by succulents (MacMahon 2000). Above the semi-deserts and steppes, open Pinyon-Juniper woodlands of cembroid pines (*Pinus monophylla, P. edulis*) and *Juniperus osteosperma* are widespread throughout the region. Depending on moisture conditions, different types of conifer forests occur higher up: while yellow pine forests with *Pinus ponderosa* occur on the leeward side of the Sierra Nevada as well as in the Eastern part of the Intermountain Region, the drier Intermountain Region is characterized by open woodlands with *Pinus longaeva* and/or *P. flexilis* at upper elevations. Between the two woodland belts, a narrow conifer-free zone (so-called "treeless balds", see Fig. 3 and Billings & Mark 1957, Höllermann 1973) might be present as a result of floral history. Further east, this gap is closed by mixed conifer forests, becoming successively "darker" eastward with the co-occurrence of Douglas fir (*Pseudotsuga menziesii*), spruce (*Picea engelmannii*) and fir (*Abies lasiocarpa*), as the amount of summer rain increases. In the Eastern part the latter two form even pure spruce-fir forest. Above tree-line, which is located at approximately 3400 m a. s. l., loose talus communities are widespread on steep slopes with patchy forb-rich turfs, low (sub-) shrubs and cushion plants.

Three general types of the arrangement of altitudinal belts can be recognized within the region (Fig. 4, see also Richter 2000, Fickert 2006). The homogeneous type showing a high symmetry between the opposing declivities is widespread within the Intermountain Region. Similar climatic conditions on both flanks are reflected by high taxonomic similarities between opposite sides. In contrast, rainshadow effects are reflected by a more or less pronounced asymmetry of vegetation belts between western and eastern slopes (e. g. Sierra Nevada). Intermediate conditions with shifted boundaries or additional vegetation belts inserted on one side, but missing on the other, often are caused by differences in radiation input (north vs. south aspect) or occur where differences between two opposing sides are minor.

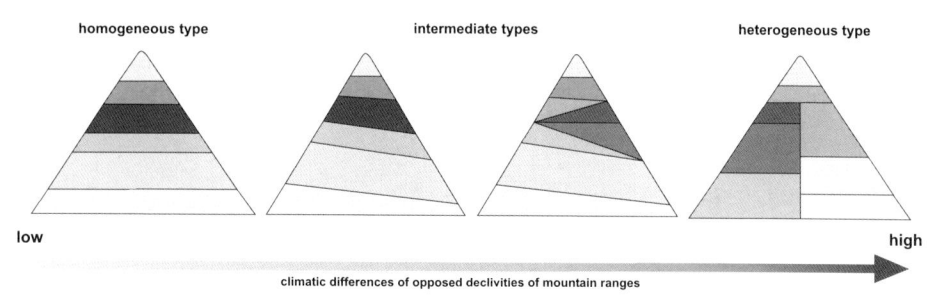

Fig. 4 Schematic diagram showing general types of arrangement of elevational belts in mountain systems (modified from Richter 2000).

3.2 Geodiversity as Trigger for Plant Diversity

As already mentioned, the American Southwest is one of the most species rich "dry regions" of the world. With up to 2,000 taxa per 10,000 km² (Barthlott et al. 2007) and 3,000 per ecoregion (Kier et al. 2005), the total number of species in the Southwest surpasses that of other dry regions of the world by far. This high gamma-diversity is even more astonishing as the alpha-diversity in the South-Western USA is not particularly outstanding. Fig. 5 displays the species numbers of 69 test sites (500 m² each) in zonal vegetation units all over the Western United States at different altitudes (data from Fickert 2006). With few exceptions, the species numbers per sample range between 15 und 35. These values are well surpassed in tropical communities and even extensively used meadow communities of the temperate zone generally harbor more different taxa (unpublished data).

Besides the moderate alpha-diversity, Fig. 5 highlights another scale dependent diversity aspect of the area. While there are strong correlations between climatic parameters (such as potential evapotranspiration, mean annual temperature, length of growing season, etc.) and the number of species documented for North American ecoregions (e. g. Mutke & Barthlott 2005), there is no such coherence apparent for the alpha-diversity of the 69 test sites in Fig. 5, neither for temperature, nor for hygric measures of annual precipitation and the number of arid months per year.

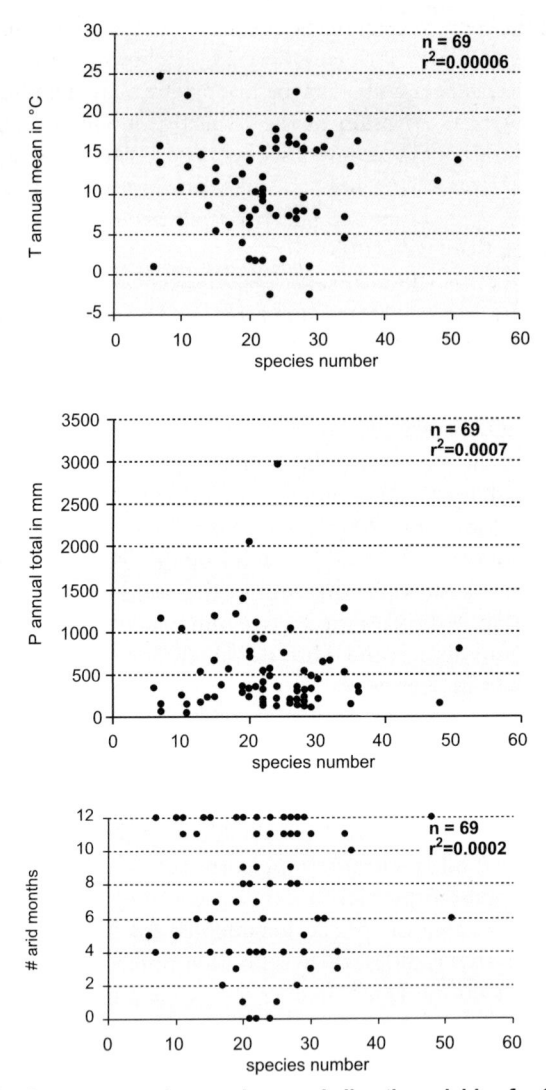

Fig. 5 Correlation between species numbers and climatic variables for 69 test sites in zonal vegetation units (i. e. not riparian, burned, anthropgenic influenced, etc., 500 m² each) all over the Western United States (North to South, West to East and from the lowland to high elevation). The different location of the samples is reflected by the different climatic values on the ordinate (data from FICKERT 2006).

In contrast, a remarkably common feature in many different groups of organisms within the area is the high beta-diversity (see TUELLER et al. 1991 for plant communities, MAC NALLY et al. 2004 for birds and butterflies, and DAVIS (2005) for mammals). The high beta-diversity is clearly dependent on the high geodiversity of the area. Due to the geologic-tectonic development of the area, bedrock of different age and chemism is exposed in close proximity, which creates a wide variety of soils and topography, thus favoring high species turnover rates from one place to another. Even more important is the complex environmental gradient of elevation. Regarding plants, the larger the vertical extension, the more vegetation belts can be developed as a result of the changing hygro-thermic conditions with elevation. Fig. 6 shows a strong relation between the richness of species, genera and – maybe not as obvious – families in local mountain floras and the vertical extension of the respective mountain range (GRÜNINGER & FICKERT 2003). Similar patterns are described by BOWERS & MCLAUGHLIN (1982) for local floras from Arizona, New Mexico and Western Texas. This correlation is even higher on heterogeneous mountain ranges with their higher amount of different vegetation belts developed (RICHTER 2000, GRÜNINGER & FICKERT 2003, FICKERT 2006). Fig. 6 once more shows the effect of substrate on diversity. The remarkable difference in species number between the White Mountains and the Spring Mountains is because the former range is petrographically rather varied, while the latter consists exclusively of limestone.

Intermediate mountain ranges (Fig. 4) also act somewhat intermediate, concerning the various diversity measures, as they have some aspects in common with the homogeneous types and some with the heterogeneous. While alpha-diversity of samples is generally independent of the mountain range type, beta-diversity between opposing aspects may differ. If shifted belts or additional belts occur on one side, species turnover rates are similar to heterogeneous mountain ranges, where different units occur (compare Fig. 4). Where the same belts are met, beta-diversity is more like in the case of homogeneous ones. Gamma-diversity, finally, is not affected by shifted belts; however, added units on one side may cause an increase in total species number, if those species are not present in the range otherwise.

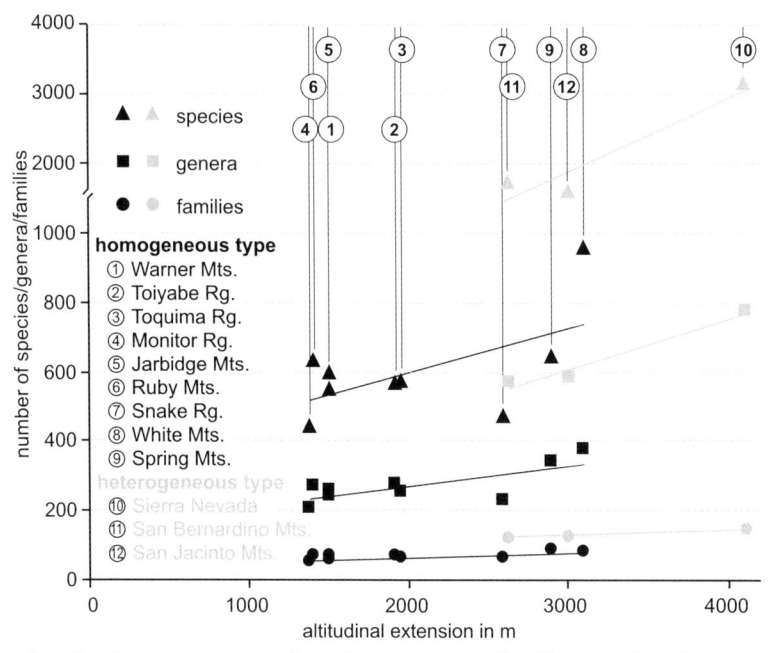

Fig. 6   Relation between number of species, genera and families and elevational range in 12 local floras/datasets (data from CLOKEY 1951, MOREFIELD et al. 1988, CHARLET 2001 and *Jepson Flora Project* w/o. y.).

3.3 The Role of Floral History

In our search for evidence for the high gamma-diversity in the Southwest of the US we will continue with the aspects of floral history, which are of major concern for the actual composition of plants there as elsewhere in the world. The evolution of the South-Western floras can be traced back to Paleozoic time, even if the contribution to present-day floras of these long-gone elements is rather indirect than direct. Direct contributions, though, had processes since Tertiary time, in particular, long distance migration of so-called "geofloras" and the in-situ evolution (i. e. mutation, natural selection etc.), two phenomena that commonly occur simultaneously, as the gene pool of migrating populations most likely will be altered by adaptations and natural selection (Tidewell et al. 1972).

Three such geofloras existed in early Tertiary time: a broad-leafed, mesophytic, evergreen Neotropical-Tertiary flora in the South; an Arcto-Tertiary flora consisting of temperate conifer and mixed deciduous trees/shrubs in the North; and a micro- and scleropyllous Madro-Tertiary flora (named after the Sierra Madre in Mexico) evolving on drier sites within the Neotropical-Tertiary Geoflora (Tidwell et al. 1972, Axelrod 1950, Brown 1994). Climatic changes created by the Sierran uplift during the Tertiary (most pronounced since late Miocene/early Pliocene, but continuing into the Pleistocene) strongly influenced the present-day vegetation. The increasing aridity favored the spread of the Madro-Tertiary flora, and the evergreen chaparral, the microphyllous desert scrub as well as the drought-deciduous thorn scrub so wide spread in the Southwest today, all are remnants or derivatives of these Madro-Tertiary elements. The Neotropical-Tertiary Geoflora was simultaneously pushed south, while Arcto-Tertiary elements (conifers and deciduous hardwood taxa) retreated north and up to higher elevations in the mountains.

The final development and molding of the present day floristic arrangement of the Southwest occurred during the cyclic climatic changes in the Pleistocene. During warmer and/or drier interglacial times, plants were able to invade the area from the south contributing southern elements to the present lowland flora (Reveal 1979). During cold cycles, in contrast, the major mountain ranges served as migratory pathways for arctic and boreal plants recessing southwards from the advancing continental ice shields (Axelrod & Raven 1985, Billings 2000). Thus, many species of northern origin reached far south and today are restricted to the higher elevations of the Rocky Mountains, the Sierra Nevada, and (to a lesser extent) the Intermountain Region. In these numerous isolated mountain ranges, a cyclic up- and downward displacement of plants apparently was more important, documented in the effect that numerous (sub-)alpine plants of Intermountain mountain ranges derived from lowland (= desert) species with the ability of adaptation (Weber 1965, Axelrod & Raven 1985).

Due to their segregation, these mountain ranges stimulated the application of the island biogeography theory of MacArthur & Wilson (1967) to explain the present day composition of organisms (Brown 1971, 1978, Johnson 1975, 1978, Harper et al. 1978, Billings 1978, Thompson & Mead 1982, Wells 1983). According to this model, the actual species number is controlled by size of the island and distance to mainland source areas, because these two factors trigger the processes of extinction (on smaller islands the chance of falling below a critical value in effective population size is higher than on larger islands) and immigration (the nearer, the easier), respectively. Applied to the Intermountain region, the dry interior valleys collectively are considered as "ocean", the two main cordilleras (Sierra-Cascades-Axis and Rocky Mountains) as "mainland", and the isolated mountain ranges in between are the "islands", therefore characterized by (1) lower species numbers than the adjoining "mainlands", (2) decreasing species numbers with distance to the "mainland" and (3) a higher rate of endemism due to isolation.

All the earlier contributions mentioned above found a strong floristic relationships between the Intermountain Region and the Rocky Mountains, because climatic conditions in the Intermountain Region are more similar to the Rocky Mountains than to the Pacific Cordillera, thus eastern species should have had a clear advantage in colonizing. However, more recent studies by CHARLET (1991, 1995) using a broader dataset and more complete species lists, showed that neither the Eastern nor the Western Cordillera dominate in contributing to the upland flora in-between. Furthermore, on a coarse scale of resolution, the higher elevations of the two Cordilleras and the mountain ranges in between actually possess a single flora. In fact, the alpine flora of the Great Basin is distinct only on fine scale, but not due to a high rate of endemism but because of the relative mixture of western and eastern species (CHARLET 1991) and in many genera taxa are different only on subspecific rank.

This mixing can be found in many groups of organisms, but shall be illustrated here by the help of the conifers. Fig. 7 shows the distribution of conifer taxa from both mainland-source areas within the Intermountain region (CHARLET 1995). While longitude is a poor predictor for the total number of conifers present per area, a strong decrease of taxa with "Pacific" or "Rocky Mountain" origin becomes clear with distance to the respective source area. Members of both conifer groups experience successively more uncomfortable growing condition with distance to their respective core area.

Long lasting isolation favored speciation by adaptation, mutation and natural selection, often producing geographical vicariances. For example, the close relatives *Pinus balfouriana*, *P. longaeva*, and *P. aristata* had joint Cenozoic ancestors (AXELROD 1986) but today occur in clearly confined regions with specific climatic conditions. In other cases even closer related representatives replace each other. For instance, the western taxa *Abies concolor* var. *lowiana*, *Pinus ponderosa* var. *ponderosa*, *Pseudotsuga menziesii* var. *menziesii*, correspond to the eastern *Abies concolor* var. *concolor*, *Pinus ponderosa* var. *scopulorum*, *Pseudotsuga menziesii* var. *glauca* in comparable habitats.

The above mentioned "treeless balds" appear to be a result of floral history, too. While edaphic or climatological reasons fail to explain this phenomena, the lack of adequate conifer species (i. e. cold and dry tolerant species), which simply not (yet) invaded and/or adapted to the prevalent ecological conditions of the area is more likely (BILLINGS & MARK 1957, HÖLLERMANN 1973). Given enough time, this altitudinal gap might be closed by successively more drought adapted taxa invading either from the east or west.

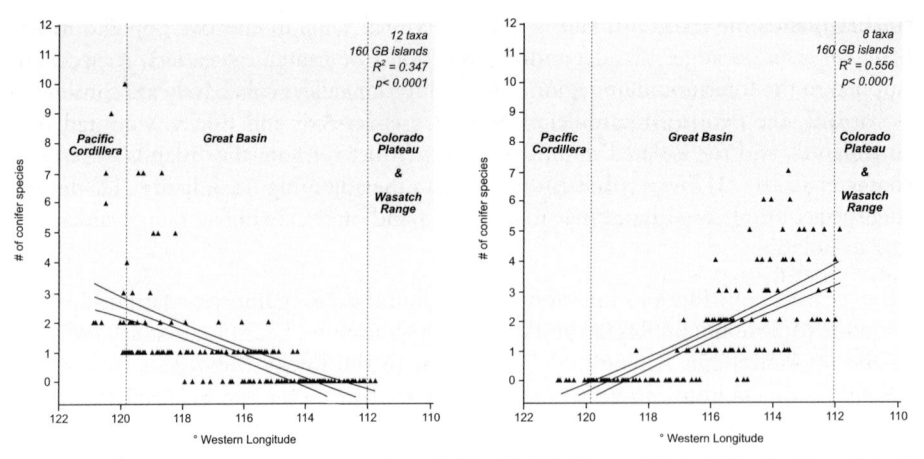

Fig. 7 Regression between the number of "Pacific" (left, n = 12) and "Rocky Mountains" conifers (right, n = 8) and longitude for 160 Great Basin mountain ranges (adapted from Charlet 1995).

3.4 Zooming In – Examples on Finer Scales from the Snake Range (Eastern Nevada)

The following subchapters present some exemplary observations of plant diversity patterns in relation to finer scales. Being fully aware of the controversial discussion on scale terminology among landscape ecologists and not wanting to add to the variety in terms, we follow the nomenclature of Willis & Whittaker (2002) in the subsequent paragraphs: Here, the *landscape scale* deals with species richness between communities within a "landscape" (i. e. mountain range) or along environmental gradients such as altitude, whereas the *local scale* is applied to species richness within single communities, habitat patches, or along short environmental gradients (i. e. ecotonal sequences across adjacent vegetation belts). As case study we have chosen the Snake Range in eastern Nevada. Clearly, the presented examples can only give some first ideas about the diversity patterns of that area and their triggers (for more details see Grüninger 2005).

The Snake Range is located in the Great Basin between 39° 30' N and 38° 30' N and is home to the Great Basin National Park established in 1986. It represents the easternmost high mountain range of Nevada, and cumulates in the 3982 m a. s. l. high Wheeler Peak, the second highest point in Nevada. As tilted fault block, the western flank of the Snake Range is steep, whereas the eastern slope is more gradual. Deeply incised streams dissect this side of the mountain and create a great variation in aspect, water availability and habitats.

As typical Great Basin mountain range the Snake Range shows the common altitudinal zonation of the area. The steppe vegetation of the foothill zone is dominated by *Artemisia tridentata*. This community is altered by the ranching history, with a high amount of introduced annual grasses successfully competing for the available water resources and thus suppressing native perennials. Above 1800 m a. s. l. the lower montane Pinyon-Juniper woodland with *Pinus monophylla* and *Juniperus osteosperma* gradually replaces the steppes. On edaphically dry sites large patches of mountain mahogany (*Cercocarpus ledifolius*) prevail, which also extend into the adjacent upper montane conifer forest. Here, *Abies concolor* var. *concolor* dominates in association with *Pseudotsuga menziesii* var. *glauca*, *Picea engelmannii*, and *Pinus flexilis*. *Populus tremuloides* is common on mesic as well as on disturbed sites in mid-successional stages and is also associated with *P. flexilis*, *P. engelmannii* in the subalpine forest up to timberline at ~3400 m a. s. l. Above, the coarse and highly mobile weathered quartzite of the summit area allows only small patches of vegetation or hardy pioneer species in the alpine zone.

3.4.1 Landscape Scale

The locality and elevational range of the Snake Range imply a steep and complex environmental gradient from the foothill region to the top, reflected in the flora and vegetation of the range. A total of 479 higher plant species is known from the Snake Range, indicating a comparatively low gamma-diversity (Fig. 6). This might probably be caused by the lower altitudinal range and the rather homogeneous petrography in contrast to other more diverse mountain ranges like the White Mountains, but another feature seems to be more important: In the Snake Range, the broad, rather species-poor alpine and subnival belts contribute much to the total vertical extension, while the generally highly diverse belts at lower elevations are underrepresented due to the height level of the valley floors above 1500 m a. s. l. (see also Chapter 3.2 and Fig. 6).

Fig. 8 shows the total number of species in vertical steps of 100 m as well as the respective life form fractions. Highest alpha-diversity is met at mid-elevations at the ecotonal

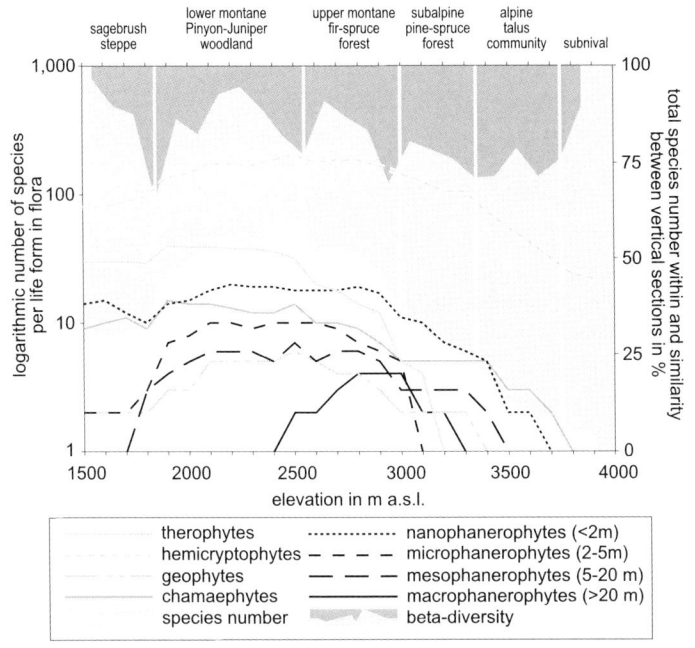

Fig. 8   Number of vascular plant species (in % of total flora), life form fractions per 100 m elevation, and similarity (Jaccard-Index) between adjacent elevation levels along the altitudinal gradient in the Snake Range (data from GRÜNINGER 2005).

sequence between lower montane woodland and upper montane forest, where the temperate climate and sufficient precipitation generally allow for favorable growing conditions. Different canopy species with distinct habitus, functional types, and stand densities create steep resource gradients and high niche diversity for understory species. Consequently, hemicryptophytes, geophytes, and nanophanerophytes also reach their peak values in these altitudes. The high percentage of therophytes in the lower elevations is of course the result of their drought avoidance strategy, but is considerably diminished by high abundance of the competitive *Bromus tectorum* (cheat grass), which was introduced to the region by livestock farming (*FICMNEW* 1997).

The between-habitat diversity of adjoining altitudinal levels (Fig. 8) in flora is generally low within the distinct vegetation belts. Unsurprisingly, turnover is greatest within ecotonal zones (see Chapter 3.4.3). Highest dissimilarity is reached at the transition between shrub steppe and the lower montane woodland, where also the change in formation structure is severe. A remarkable turnover is also recognized between the upper montane and subalpine forests. Here, next to the absence of many tree species from the lower elevations, a distinct floristic change in the understory occurs, as the radiation input near the ground increases in openings and parklands. Furthermore, depths and duration of snow cover increases and is expected to strongly alter species composition (for details see GRÜNINGER 2005 and FICKERT 2006).

The flora data is not appropriate to reveal the floral heterogeneity within the distinct vegetation belts. In Fig. 9, the species-area relations (e. g. KREBS 1999) derived from

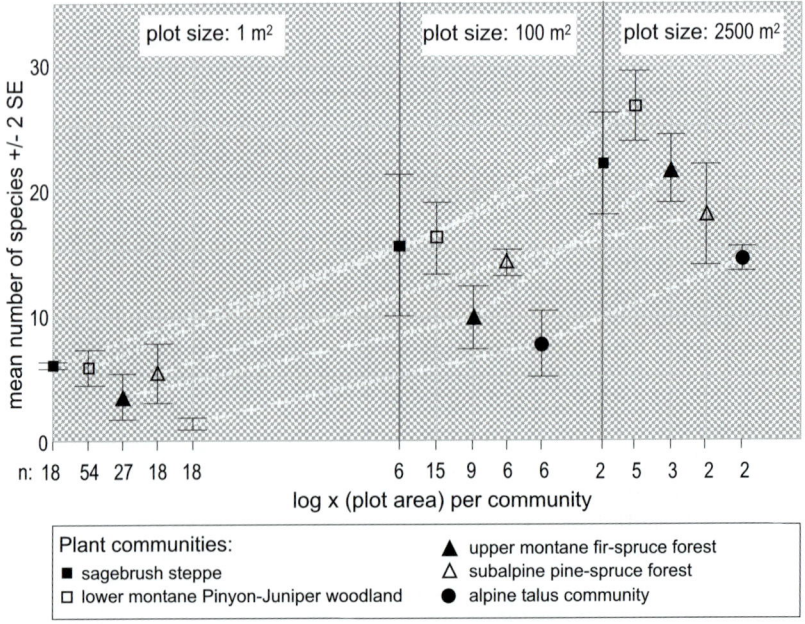

Fig. 9   Species-area relations (expressed as semi-log relation), drawn for the mean species numbers of three nested plot sizes in each zonal community of the Snake Range; for convenience only, trends are indicated by dashed lines (modified from GRÜNINGER 2005).

three nested plot sizes are due to the few sample sizes not statistically sound, but nevertheless hint at differences between the communities: high alpha-diversity is given for the 1m²-plot sizes of the steppes and lower montane woodlands, but the expected sharp increase in richness with sample size (regarding the high forb diversity in flora) is again lessened by the introduced cheat grass. In contrast, the complex structure of the upper montane forest is reflected very well in the graph. Mean species diversity is low in small and mid-sized plots and only increases strongly towards the biggest plots, comprising different tree species (see above) and the distinct associated undergrowth. The subalpine forest is characterized by higher homogeneity, as reflected by the comparatively balanced alpha-diversity values. Within talus communities, plants are restricted to few sheltered habitats and the accession of new species happens rather at random with the employed standardized sample design. Nevertheless, the displayed high species numbers in flora in Fig. 8 are again confirmed for the alpine zone.

4.3.2 Local Scale

On finer scales of observation, the mountains' environmental heterogeneity are often second in importance to direct environmental variables like water, nutrients or light, changing rapidly along gradients of different extension and across ecotonal sequences (PAUSAS & AUSTIN 2001). Furthermore, the fundamental basis of biotic and abiotic factors is modified by different dynamic processes in vegetation: especially disturbances play an important role for the gamma-diversity of the plant communities, and are a primary source for the prevailing spatial and temporal habitat complexity, of course not only in high mountains but in most ecosystems of the world in general (WHITE et al. 2000). However, due to their steep complex elevational gradients, the disturbance regimes in high mountains are a matter of high diversity itself.

An example for the change in structure and richness on disturbed sites is given in Fig. 10. Again, species-area relations are drawn for an old growth forest and an adjoining disturbed site within the upper montane forest at 2850 m a. s. l. The distance between the two plots is barely 200 m, yet they show striking differences in species composition, alpha-diversity and dominance structure.

The mature stand is characterized by the conifers *Abies concolor* var. *concolor*, *Pinus flexilis*, *P. ponderosa* var. *scopulorum*, *Pseudotsuga menziesii* var. *glauca*, *Juniperus communis* and *Picea engelmannii*. The disturbed site is a highly diverse, open community dominated by grasses, herbs and shrubs, constantly influenced by grazing and browsing wildlife and probably initialized by wildfire. With 47 species per 2500 m² (Fig. 10), the sample showed the highest alpha-diversity of all samples in the Snake Range (GRÜNINGER 2005). The high increase in richness from each plot size to the next is generated mainly by hemicryptophytic species and indicated a strong heterogeneity within the herb layer. The patchy distribution of trees, shrubs and especially herbs and the therefore strong increase in alpha-diversity is clearly a result of the constant habitat alteration by browsing and grazing wildlife, keeping the community open in diminishing biomass, fertilizing the soil and enriching the seed pool. In the late successional stage of the sample, direct influences of the suggested preceding wildfire are not pronounced anymore, but obviously perennials still find more appropriate niches than without the disturbance impact.

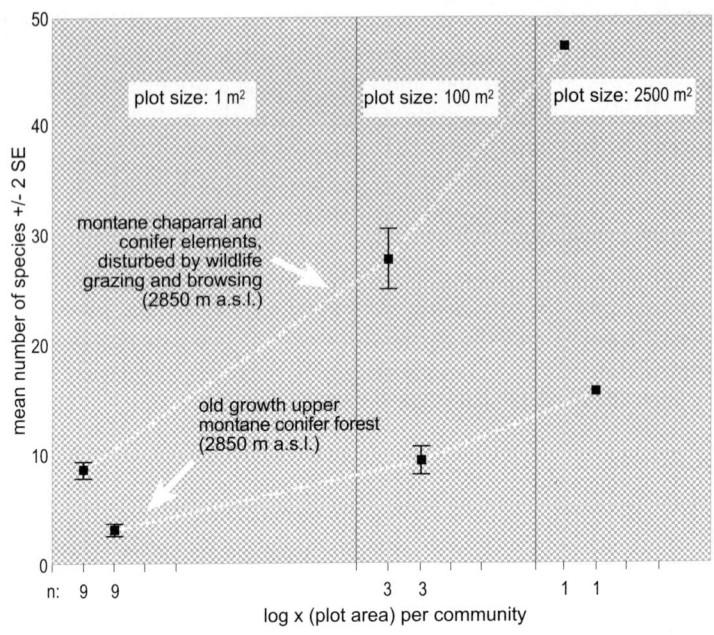

Fig. 10 Species-area relations (expressed as semi-log relation), drawn for the mean species numbers of three nested plot sizes in an old growth and a disturbed site at 2850 m a. s. l. in the Snake Range; for convenience only, trends are indicated by dashed lines (modified from Grüninger 2005).

3.4.3 Where Change Happens – Ecotones and Diversity Patterns

The presented observations document various diversity turnover situations between discrete sites or whole communities. They fail, however, in answering the question about where and how this turnover happens. Independent of scale and character but dependent on perception, distinct entities are separated or connected by transitions between them. On coarser scales, these ecotones can appear as borders between discrete units, whereas on finer scales they may represent highly diverse, sensitive and extremely flexible transition zones between areas with greater continuity in species composition and vegetation structure (e. g. Holland & Risser 1991).

High mountain regions with their increased geodiversity, their compact sequences of different vegetation belts, their diverse disturbance regimes and their varied plant communities are especially rich in ecotones. Here, species from adjacent communities intermingle and may as well be associated with highly specialized "ecotonal" species, profiting from the special habitats along the ecotones (Di Castri & Hansen 1992). But not only species composition and species richness change across an ecotonal sequence, other diversity parameters such as evenness or life form composition might change as well. Also, these diversity parameters can show significant turnover rates at different locations along transects. Fig. 11 gives an idea of these complex changes along a 200 m-transect crossing the ecotone between the upper montane conifer forest (2575 m a. s. l.) and the lower montane Pinyon-Juniper woodland (2490 m a. s. l.). As described in chapter 3.4.1, these elevations show the highest alpha-diversity in regard to vascular

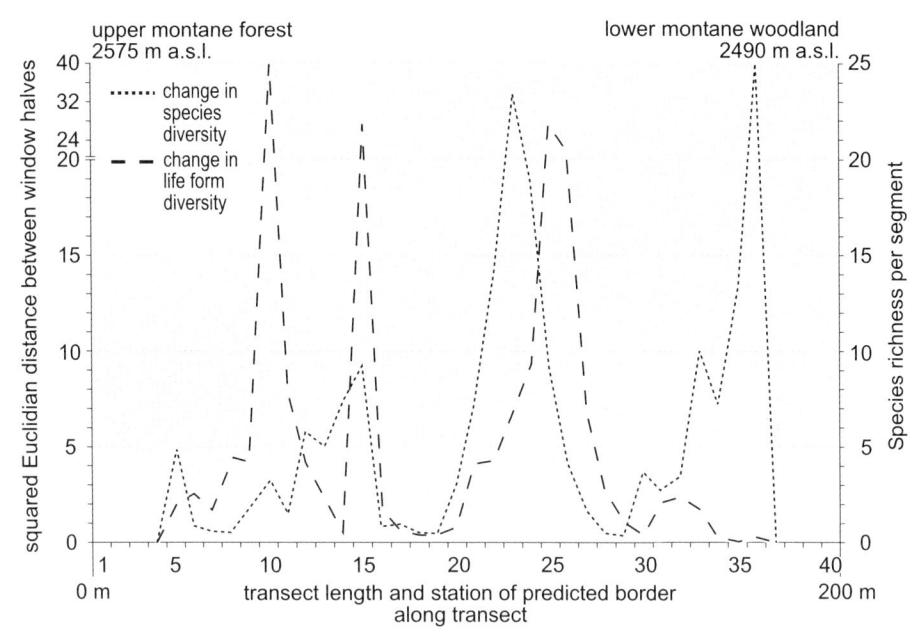

Fig. 11 Predicted borders for change in species and life forms per segment (5 m length by 2 m width, considering respective ground cover values) along the ecotonal sequence from the upper montane to lower montane forest communities in the Snake Range using SMWDA ("Split Moving Window Distance Analysis", window width: 8, processed using then first two PC's; for details see e. g. CORNELIUS & REYNOLDS 1991). Note: the peak at station 36 at the end of the transect is due to its position statistically not trustworthy.

plants (Fig. 8) and the beta-turnover between the adjoining plant communities is high. Throughout the 5 by 2 m transect segments, species numbers are relatively unstable and without a clear trend. Nevertheless, changes occur regarding quantity values of the plant species and the assigned life forms (Fig. 11). Regarding plant species richness, the most significant turnover is indicated from stations 23 to 25 and life form composition also indicates high dissimilarities at this position (25 and 26), as the dominant macrophanerophytic *Abies concolor* var. *concolor* of the upper montane is replaced by the dominant mesophanerophytic *Pinus monophylla* of the lower montane community. In contrast, the obvious change in life forms indicated at station 10 is not reflected in the species values; this pattern is apparently generated by small, but overall cumulating amounts of various nanophanerophytic shrubs in this sequence of the ecotone, which profit from the patchiness of the overstory species and the thereby increased niche diversity due to higher water availability and radiation input.

4 Plant Diversity in Mountain Systems – Still Barely Scratching the Surface

As shown in the previous chapters, plant diversity patterns are triggered, observable and explainable on different scales of observation and – on the whole – are therefore not easy to assess. In a review on "*The unified neutral theory of biodiversity and biogeography*" by HUBBELL (2001), SILANDER (2002) compares the seek for an ultimate explanation in biodiversity patterns with *Douglas Adams "The Hitchhiker's Guide to the Galaxy: the Ultimate Answer"*; Here, the computer *Deep Thought* is questioned about the ultimate answer to "Life, the Universe and Everything", and, after a tremendous amount of time thinking concludes that the answer is 42 – amazingly, the same "fundamental biodiversity number" given by HUBBELL (2001) for typical tropical rainforest systems in Panama.

The author's skepticism towards such approaches seems coherent, and caution regarding this matter has to be exercised for extratropical ecosystems like in the area covered here just as for tropical systems. Even if single processes behind its diversity are not unique but can be recognized all over the world, the interaction and interrelation of geologic history, tectonic features, topography, superimposed by the atmospheric circulation patterns and their mesoclimatic modifications, as well as floral history and dynamic processes create an exclusive situation in the rather dry Soutwestern United States nonetheless. Employing again just species numbers, the ecoregion approach by KIER et al. (2005) shows comparable species rich "dryland"-situations for the ecoregions in the vicinity of the Somali peninsular (Fig. 1) with its high-relief and diverse hygrothermic conditions. Consulting the grid-cell approach by BARTHLOTT et al. (2007), the exclusiveness of the region becomes even more pronounced. With 10,000 km², the resolution obeys the necessities of a global approach and, thereby strongly enhances the geodiversity per grid-cell, which is derived from the above mentioned factorial complex. Due to this setting, flora elements from the complete elevational gradient as well as of different habitat types contribute species to the same grid-cell and thus create the high gamma-diversity of the area.

References

AUSTIN, M. P. (1999): The potential contribution of vegetation ecology to biodiversity research. Ecography 22: 465–484.
AXELROD, I. D. (1986): Cenozoic history of some western American pines. Annals of the Missouri Botanical Garden 73: 301–306.
AXELROD, I. D. 1950. The evolution of desert vegetation in western North America. Publication of the Carnegie Institution of Washington 590: 215–306.
AXELROD, I. D.; P. H. RAVEN (1985): Origins of the Cordilleran flora. Journal of Biogeography 12: 21–47.
BARBOUR, M. G.; J. H. BURK; W. D. PITTS (1999): Terrestrial Plant Ecology. Menlo Park (California).
BARTHLOTT, W.; A. HOSTERT; G. KIER; W. KÜPER; H. KREFT; J. MUTKE; M. D. RAFIQPOOR; J. H. SOMMER (2007): Geographic patterns of vascular plant diversity at continental to global scales. Erdkunde 61(4): 305–315.
BARTHLOTT, W.; M. WINIGER (eds.) (1998): Biodiversity – a challenge for development research and policy. Berlin, Heidelberg.
BARTHLOTT, W.; N. BIEDINGER; G. BRAUN; F. FEIG; G. KIER; J. MUTKE (1999): Terminological and methodological aspects of the mapping and analysis of global biodiversity. Acta Botanica Fennica 162: 103–110.
BARTHLOTT, W.; J. MUTKE; M. D. RAFIQPOOR; G. KIER; H. KREFT (2005): Global centres of vascular plant diversity. Nova Acta Leopoldina 342: 61–83.

Barthlott, W.; W. Lauer; A. Placke (1996): Global distribution of species diversity in vascular plants: towards a world map of phytodiversiy. Erdkunde 50: 317–327.

Beierkuhnlein, C. (1999): Räumliche Muster der Biodiversität in nordbayerischen Landschaften. Habilitation Thesis in Biogeography, University Bayreuth, (unpublished).

Billings, W. D.; A. F. Mark (1957): Factors involved in the persistence of montane treeless balds. Ecology 38: 140–142.

Billings, W. D. (1978): Alpine phytogeography across the Great Basin. Great Basin Naturalist Memoirs 2: 105–117.

Billings, W. D. (2000): Alpine Vegetation. In: M. G. Barbour; W. D. Billings (eds.): North American Terrestrial Vegetation, Cambridge (UK): 537–572.

Bowers, J. E.; S. P. McLaughlin (1982): Plant species diversity in Arizona. Madroño 29: 227–233.

Brown, D. E. (1994): Biotic Communities: Southwestern United States and Northwestern Mexico. Salt Lake City.

Brown, J. H. (1971): Mammals on mountaintops: nonequilibrium insular biogeography. American Naturalist 105: 467–478.

Brown, J. H. (1978): The Theory of insular biogeography and the distribution of boreal birds and mammals. Great Basin Naturalist Memoirs 2: 209–277.

Bryson, R. A.; F. K. Hare (1974): The Climates of North America, World Survey of Climatology, Vol. 11, Elsevier Scientific Publishing Company, Amsterdam, London, New York.

Charlet, D. A. (1991): Relationships of the Great Basin Alpine Flora: a quantitative analysis. Master Thesis in Biology, University of Nevada, Reno (unpublished).

Charlet, D. A. (1995): Great Basin montane and subalpine conifer diversity: dispersal or extinction pattern? PhD Thesis in Biology, University of Nevada, Reno (unpublished).

Charlet, D. A. (2001): Plants of the Great Basin, www.brrc.unr.edu/data/plants

Clokey, I.W. (1951): Flora of the Charleston Mountains, Clark County, Nevada. Berkeley and Los Angeles.

Cornelius, J. M.; J. F. Reynolds (1991): On determining the statistical significance of discontinuities within ordered ecological data. Ecology 72/6: 2057–2070.

Crawley, M. J.; J. E. Harral (2001): Scale dependence in plant biodiversity. Science 291: 864–868.

Davis, E. B. (2005): Mammalian beta diversity in the Great Basin, western USA: palaeontological data suggest deep origin of modern macroecological structure. In: Global Ecology and Biogeography, 14: 479–490.

Di Castri, F.; A. J. Hansen (1992): The environment and development crisis as determinants of Landscape dynamics. In: Hansen, A. J.; F. Di Castri (eds.): Landscape boundaries; Consequences for biotic diversity and ecological flows. Ecological Studies 92: 3–18.

Fickert, Th. (2006): Phytogeographische Studien als Mittel zur Klimaableitung in Hochgebirgen – Eine Fallstudie im Südwesten der USA. Passauer Schriften zur Geographie 22.

FICMNEW (1997): Invasive plants – changing the landscape of America. Washington, D. C.

Gaston, K. J. (1998): Biodiversity – The road to an atlas. Progress in Physical Geography 22/2: 269–281.

GRÜNINGER, F.; TH. FICKERT (2003): Revealing diversity patterns of vascular plants and their causes in semiarid high mountain regions – a top down approach for Great Basin mountain ranges, USA. Erdkunde 57/3: 199–215.

GRÜNINGER, F. (2005): Scale dependent aspects of plant diversity in semiarid high mountain regions. An exemplary top-down approach for the Great Basin (USA). Passauer Schriften zur Geographie 21.

HARPER, K. T.; D. C. FREEMAN; W. K. OSTLER; L. C. KILKOFF (1978): The flora of Great Basin mountain ranges: diversity, sources and dispersal ecology. Great Basin Naturalist Memoirs 2: 81–103.

HOLLAND, M. M.; P. G. RISSER (1991): The role of landscape boundaries in the management and restoration of changing environments. In: HOLLAND, M. M.; P. G. RISSER; R. J. NAIMAN (eds.): Ecotones: The role of landscape boundaries in the management and restoration of changing environments. New York, London: 1–7.

HÖLLERMANN, P. (1973): Some Aspects of the Geoecology of the Basin and Range Province (California Section). – Arctic and Alpine Research 5 (3), S. A85–A98.

HUBBELL, S. P. (2001): The unified theory of biodiversity and biogeography. Monographs in Population Biology 32, Princeton University Press. Princeton, NJ.

HUSTON, M. A. (1994): Biological Diversity. Cambridge University Press.

Jepson Flora Project (wo.y.): http://ucjeps.berkeley.edu/interchange/index.html

JOHNSON, N. K. (1975): Controls of number of bird species on montane islands in the Great Basin. Evolution 29: 545–567.

JOHNSON, N. K. (1978): Patterns of avian geography and speciation in the Intermountain Region. Great Basin Naturalist Memoirs 2: 137–159.

KENT, M. (2005): Biogeography and macroecology In: Progress in Physical Geography 29/2: 256–264.

KIER, G., J. MUTKE; E. DINERSTEIN; T. H. RICKETTS; W. KÜPER; H. KREFT; W. BARTHLOTT (2005): Global patterns of plant diversity and floristic knowledge. Journal of Biogeography 32 (7): 1107–1116.

KREBS, C. J. (1999): Ecological methodology. Menlo Park, CA.

KREFT, H.; W. JETZ (2007): Global patterns and determinants of vascular plant diversity. Proceedings of the National Academy of Sciences (PNAS), 104: 5925–5930.

LEVIN, S. A (1992): The problem of pattern and scale in ecology. Ecology 73: 1943 – 1983.

LOWE, C. H. (1961): Biotic communities in the sub-Mogollon region of the inland Southwest. In: Journal of Arizona-Nevada Academy of Science 2: 40–49.

MAC NALLY, R.; E. FLEISHMAN; L. P. BULLUCK; C. J. BETRUS (2004): Comparative influence of spatial scale on beta diversity within regional assemblages of birds and butterflies. Journal of Biogeography 31: 917–929.

MACARTHUR, R. H. (1965): Patterns of species diversity. Biological Review 40: 510–533.

MACARTHUR, R. H.; E. O. WILSON (1967): The Theory of Island Biogeography. Princeton University Press.

MACMAHON, J. A. (2000): Warm Deserts. In: M. G. BARBOUR; W. D. BILLINGS (eds.): North American Terrestrial Vegetation. Cambridge (UK): 285–322.

MCINTOSH, R. P. (1967): An index of diversity and the relation of certain concepts to diversity. Ecology 48: 392–404.

MOREFIELD, J. D.; D. W. TAYLOR; M. DEDECKER (1988): Vascular flora of the White Mountains of California and Nevada: An updated, synonymized working checklist. In: HALL, C. A.; V. DOYLE-JONES (eds.): Plant Biology of Eastern California: 310–364.

Mutke, J.; W. Barthlott (2005): Patterns of vascular plant diversity at continental to global scales. Biol. Skr. 55: 521–531.

Pausas, J. G.; M. P. Austin (2001): Patterns of plant species richness in relation to different environments: An appraisal. In: Journal of Vegetation Science 12: 153–166.

Peet, R. K. (1974): The measurement of species diversity. Annual Review of Ecology and Systematics 5: 285–307.

Powell, D. R.; H. E. Klieforth (1991): Weather and Climate. In: Hall, C. A. (ed.): Natural History of the White-Inyo Range, eastern California: 3–26.

Reveal, J. L. (1979): Biogeography of the Intermountain Region. Mentzelia 4, 1–92.

Richter, M. (2000): A hypothetical framework for testing phytodiversity in mountainous regions: the influence of airstrems and hygrothermic conditions. Phytocoenologica 30 (3/4): 519–541.

Ricklefs, R. E.; D. Schluter (1993): Species diversity in ecological communities: historical and geographical perspectives. University of Chicago Press, Chicago.

Rosenzweig, M. L. (1995): Species diversity in space and time. Cambridge University Press, Cambridge.

Sheppard, P. R.; A. C. Comrie; G. D. Packin; K. Angersbach; M. K. Hughes (2002): The climate of the US Southwest. In: Climate Research 21: 219–238.

Silander, J. A. Jr. (2002): The Ultimate answer ... is 42? Special book review of S.P. Hubbell (2001): The unified theory of biodiversity and biogeography. Journal of Biogeography 29: 299–301.

Stokes, W. L. (1986): Geology of Utah. Occasional Paper # 6 of the Utah Museum of Natural History, Salt Lake City.

Thelin, G. P.; R. J. Pike (1991): Landforms of the conterminous United States – A digital shaded-relief portrayal, USGS map I-2206 (http://wrgis.wr.usgs.gov/I-map/i2206).

Thompson, R. S.; J. I. Mead (1982): Late Quaternary environments and biogeography in the Great Basin. Quaternary Research 17: 39–55.

Tidewell, W.D.; S. R. Rushforth; D. Simper (1972): Evolution of the Floras in the Intermountain Region. In: Cronquist, A.; A. H. Holmgren; N. H. Holmgren; J. L. Reveal (1972): Intermountain Flora Vol. 1: 19–39. New York.

Tueller, P. T.; R. J. Tausch; V. Bostick (1991): Species and plant community distribution in a Mojave–Great Basin desert transition. Vegetatio 92: 133–150.

Turner, M. G.; R. H. Gardner; R. V. O'Neill (2001): Landscape Ecology. Springer, Berlin, Heidelberg, New York.

UN (1993): Convention on Biological Diversity (with annexes). Concluded at Rio de Janeiro on 5 June 1992. United Nations. Treaty series, Vol. 1760, I 30619.

Weber, W. A. (1965): Plant geography in the Southern Rocky Mountains. In: Wright, H. E.; D. G. Frey (eds.): The Quaternary of the United States, Princeton University Press, Princeton, N. J., 453–468.

Weischet, W. (1996): Regionale Klimatologie. Teil 1: Die Neue Welt. Teubner, Stuttgart.

Wells, P. V. (1983): Paleobiogeography of montane Islands in the Great Basin since the last pluvioglacial. Ecological Monographs 53/4: 341–382.

White, P. S. (1979): Pattern, processes and natural disturbance in vegetation. The Botanical Review 45: 229–299.

White, P. S.; J. Harrold; J. L. Walker; A. Jentsch (2000): Disturbance, scale and boundary in wilderness management. In: Cole, D. N.; S. F. McCool (eds.): Wilderness science in a time of change. Proceedings RMRS-P-000. USDA, FS, RMRS, Ogden, Utah.

Whittaker, R. H. (1970): Communities and ecosystems. Current concepts in biology series, Macmillan Company, New York.
Whittaker, R. H. (1972): Evolution and measurement of species diversity. Taxon 21: 213–251.
Whittaker, R. H. (1977): Evolution of species diversity in land plant communities. In: Hecht, M. K.; W. C. Steere; B. Wallace (eds.): Evolutionary Biology 10, pp. 1–67. Plenum, New York.
Willis, K. J.; R. J. Whittaker (2005): Species diversity – scale matters. In: Science 295: 1245–1248.
Wilson, E. O.; F. M. Peter (1988): Biodiversity. National Academic Press, Washington D.C.
Wilson, E. O. (1985): The biological diversity crisis: a challenge to science. Issues in Science and Technology 11(1): 22–29.
Wilson, M. V.; A. Shmida (1984): Measuring beta-diversity with presence-absence data. Journal of Ecology 72: 1055–1064.

Plant Diversity Patterns and Reindeer Pastoralism in Northern Norwegian Mountain Systems

Jörg Löffler & Roland Pape[1]

Abstract

It is likely that the change in reindeer pastoralism from subsistence to profit-oriented meat production, which was forced by a technological revolution since the 1950s and accompanied by increasing stocking rates, has left its mark on the mountain systems used as pastures. Herbivory is known to strongly influence arctic and alpine vegetation. It affects species composition, biomass, and plant architecture. Against this background, the literature on plant diversity patterns and their determination, on reindeer pastoralism in northern Norway, and on the effects on ecosystems of the grazing behaviour of herbivores is reviewed and compared to results of a case study on the environmental determination of patterns of plant diversity in mountain systems of northern Norway.

1 Introduction

1.1 Why does diversity matter?

The stimulating issue of why diversity matters was posed by Körner (2004), who listed four reasons that are particularly relevant for biodiversity in mountain systems: (i) organisms have a right to exist (the ethical argument); (ii) diverse, man-made ecosystems are treasures of society (the cultural heritage argument); (iii) the interdependence of species, diversity provides insurance against system failure by providing functional redundancy in traits (the ecological argument); and (iv) diversity provides insurance against failure in terms of the quality and quantity of desired products, ecosystem services through sustained ecosystem functioning (the economic argument).

Mountains are widely recognized as containing highly diverse and species-rich ecosystems (e. g., Diaz et al. 2003, Körner 2004). Topographic gradients condensed over short distances are capable of producing unique hotspots of biodiversity. Environmental heterogeneity is commonly expected to be correlated positively with numbers of species (Rosenzweig 1995, Körner 2004). Yet, at the same time, high mountain ecosystems are highly vulnerable to changes in the environment, for instance, slow land degradation due to human activities with all the attendant socioeconomic consequences (Messerli and Ives 1997). Furthermore, many studies suggest that mountain systems are among the most sensitive to climatic changes that occur on a global scale. Relatively small perturbations in global processes can cascade down to produce large changes in most of the interdependent patterns and processes, from the hydrological cycle to the complex fauna and flora, and the people that depend on those resources (e. g., Thompson 2000, Beniston 2003). Human welfare, and especially the water supply, depend directly or indirectly on the functional integrity of mountain ecosystems with vegetation as their key component (Körner 2004). That being so, plant diversity provides insurance

---

[1] Department of Geography, University of Bonn, D-53115 Bonn, Germany; joerg.loeffler@uni-bonn.de; pape@giub.uni-bonn.de.

against system failure by providing functional redundancy in traits (Körner 2004). The understanding and explanation of mechanisms that control diversity is critical for predicting changes in patterns of diversity that result from changes in land use or a changing climate (Walker 1995).

Patterns of spatial and temporal variations in species richness and diversity, as well as their relationship to underlying environmental variables in arctic and alpine regions, have recently attracted considerable interest from ecologists and biogeographers (Heikkinen and Neuvonen 1997, Gough et al. 2000, Moser et al. 2005). The compressed climatic gradient, sharp ecotones, and altitudinal gradients are useful for investigating patterns in species richness, given that a changing climate may lead to the migration of species (Lomolino 2001, Grytnes 2003, Fosaa 2004, Sánchez-Gonzáles and López-Mata 2005).

When focusing on the determinants of species diversity, an appropriate consideration of scaling issues is necessary (Rosenzweig 1995). In his assessment of causes of alpine plant diversity, Körner (1995) defined a set of "sieves" that act at different spatial scales. In a global context, the harsh environmental conditions that are present in arctic-alpine landscapes require plants to possess evolutionary adaptations to low temperatures and short growing seasons, which results in a small overall species pool (Körner 2003). At finer scales, environmental heterogeneity gains importance. At the regional scale, soil parent material (Molau 2003) and grazing pressure are likely to affect patterns of diversity (Löffler and Pape 2008). For instance, moderate herbivory in productive areas is assumed to increase biodiversity, whereas diversity is reduced due to herbivory in less productive habitats (Austrheim and Eriksson 2001). At the local scale, differentiations in topography and associated changes in snow cover and soil moisture are important determinants of diversity (e. g., Gould and Walker 1999, Körner 2004). Thus, at the local scale, both regional and local determinants, and their interaction, may control diversity and need to be taken into account when attempting to understand patterns of diversity (Gough et al. 2000, Holten 2003).

1.2 Reindeer Pastoralism in Northern Norway

In northern Norway, reindeer herding is restricted by law to Sámi people. Reindeer (*Rangifer tarandus* L.) require little tending, do not need shelter, and are able to find food on their own throughout the year (Bronny et al. 1985). Thus, a specialised livelihood based on reindeer husbandry could develop early and persist for many centuries. The grazing system follows complex migration patterns. Reindeer in Finnmark county migrate from more or less snow-covered lichen pastures in inner Finnmark in wintertime, when ground and arboreal lichens constitute their diet, to green summer pastures near the coast (Fig. 1, Fig. 2), where they feed mainly on herbs, grasses, and leaves of deciduous trees (Danell et al. 1994). Fall and spring pastures have a composition that lies between these extremes (Riseth et al. 2004).

Over the last few decades, the reindeer pastoralism system has experienced immense changes initialised by exogenous factors. The increasingly greater dependence on motorized vehicles, such as snowmobiles, cars, and all-terrain vehicles (ATVs), seems to have had a particular triggering effect on other economic and societal processes in

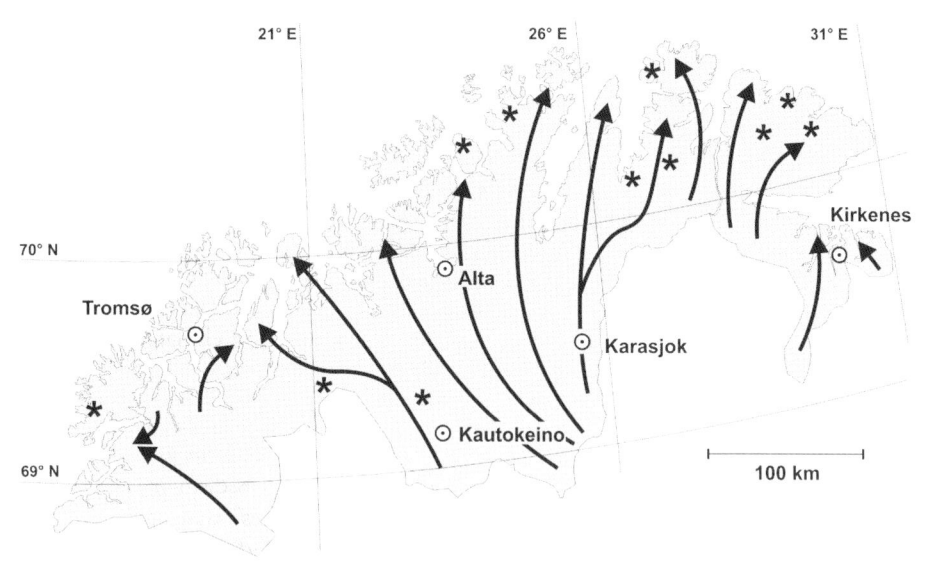

Source   Reindriftsforvaltningen 2007

Fig. 1   Map of northern Norway. Arrows indicate the spring migration of reindeer. Asterisks refer to the investigated areas of the case study.

Photo   J. LÖFFLER

Fig. 2   Reindeer herd on summer pasture

the herding society, the speed of change in which has increased. One may speak of a technological revolution in herder societies. The change from subsistence to a market economy has created great challenges for reindeer pastoralism, which is in danger of turning from a reindeer herding culture into a culture of reindeer management and the profit-oriented production of meat (BURKHARDT 2004). The need for greater production due to the high costs of modern reindeer herding has resulted in a strong increase in animal stock. The animal stock increased from about 50,000 in winter 1950 to nearly 200,000 in winter 1989. Thereafter, the number of reindeers fell to 110,000 in winter 2001 before it increased again due to good winters (less snow) to about 165,000 in 2004 (Fig. 3).

In addition to depending on suitable ranges, pastoralism affects ecosystems. Herbivores exert substantial impacts on the structure and diversity of plant communities because they decrease the abundance of preferred species (BAZELY and JEFFERIES 1986, VIRTANEN et al. 1997, BRÅTHEN and OKSANEN 2001, GRELLMANN 2002) and alter the competitive interactions between plants (MULDER and RUESS 1998, VIRTANEN 1998, OLOFSSON et al. 2002).

2 Plant Diversity and Reindeer Pastoralism

Herbivory is known to influence arctic and alpine vegetation strongly (KALELA et al. 1961, BATZLI et al. 1980). Species composition, biomass, and plant architecture are among the main components of plant communities that can be affected by herbivory (McNAUGHTON 1984). ZIMOV et al. (1995) hypothesised that large herbivores could drive a biome-shift from moss-dominated tundra into steppe-like, graminoid-dominated communities. According to their hypothesis, herbivory favours more productive plant species, improves the litter quality, increases the soil temperature by removing the isolating moss-/lichen-carpet, and consequently increases the net mineralisation rate and the productivity. Changes in vegetation due to herbivory were mostly found

Source   Reindriftsforvaltningen 2007

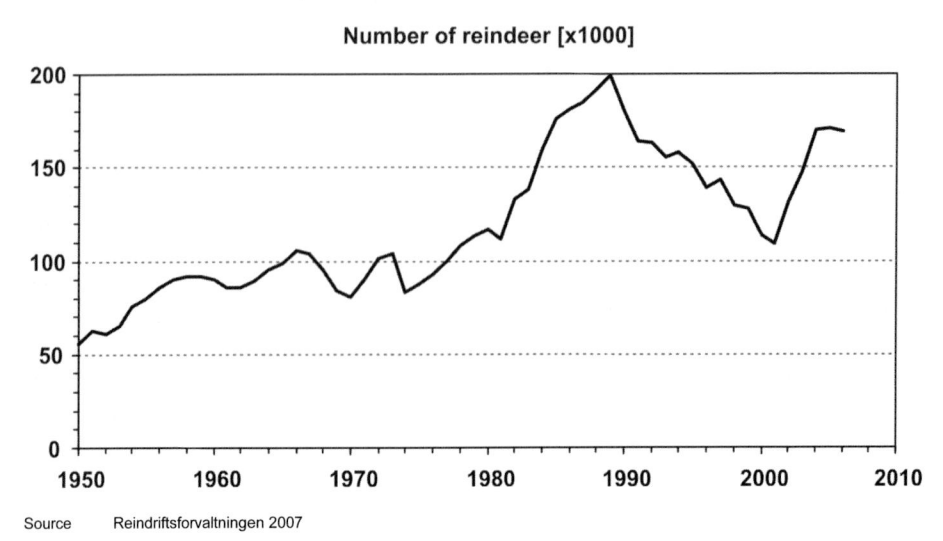

Fig. 3   Development of reindeer stock in Finnmark county

to be slow but continous, which complicates both their documentation over a period of years (Wegener and Odasz-Albrigtsen 1998) and the differentiation between effects caused by weather variation and those caused by herbivory. The importance of long-term studies was also stressed by Olofsson et al. (2004), who reported a fast decrease of dwarf-shrubs due to grazing, while the increase of graminoids took much more time. Changes in the composition and biomass of fodder plants in pastures, especially those that are most important to the herbivore, were also reported by Manseau et al. (1996), Augustine et al. (1998), Crête and Doucet (1998), Virtanen (2000), Olofsson et al. (2001, 2002, 2004), den Herder et al. (2003, 2004), and Boudreau and Payette (2004). According to Oksanen et al. (1981) and Oksanen and Oksanen (2000), in unproductive tundra systems, herbivores are more important for limiting the plant biomass than is nutrient availability for plants. The studies of Olofsson et al. (2001, 2004), in particular, support Zimov et al.'s hypothesis. Reindeer grazing was found to increase soil temperatures, litter decomposition rates, and nutrient cycling what appeared most advantageous for graminoids – they increased to the expand of woody plants. As a consequence of these changes, productivity was found to be highest in heavily grazed areas. Thus, at least for summer pastures, reindeer may enhance their own food resources and may receive positive feedback to their food supply. However, the opposite outcome of grazing has been observed in an arctic tundra heath, where herbivory was practised mainly during the winter (Grellmann 2002, Stark and Grellmann 2002). Therefore, the presence of reindeer in tundra ecosystems may have mixed effects on nutrient cycling. Reindeer's effects on ecosystem processes may be either positive or negative, depending entirely on the temporal and spatial patterns of their grazing (Olofsson et al. 2004).

Grazing is also considered to be an important factor for plant species diversity, although it is difficult to generalise about the impact (Crawley 1997, Olff and Ritchie 1998). The effects of herbivory on species diversity depend on the type of grazer, the intensity and frequency of grazing (Hobbs and Huenneke 1992), and habitat characteristics, such as productivity (Huston 1979, Virtanen 1998, Proulx and Mazumder 1998), climate, and evolutionary history (Milchunas et al. 1988). Olff and Ritchie (1998) found that large herbivores increase diversity quite consistently except at artificially high densities, whereas in the conceptual model of Austrheim and Eriksson (2001) the effects on vegetation structure and diversity at each local patch were related to grazing frequency. Grazing imposes a local shift towards a modified plant community at a particular time. If grazing continues at that particular site, the shift will become permanent; otherwise, each site will tend to recover as the grazers move to other sites. Thus, up to a certain grazing frequency (and associated intensity), the impact of herbivores is likely to increase the heterogeneity, and therefore diversity, of the species present in the landscape because each patch will be at a different stage of recovery. Regarding productivity, plant species richness, is according to Proulx and Mazumder (1998), generally favoured by grazing in productive communities. In less productive communities, grazing that is at least moderate also favours diversity, whereas heavy grazing reduces diversity. In communities in which productivity is extremely low (barren heath, extreme snow-beds), grazing could have a negative effect on plant species richness. As stated by Milchunas et al. (1988), the reaction of plant communities to different grazing pressures depends on the history of grazing. In a community that has adapted to moderate grazing pressure, both very high and very low grazing pressure can be considered as disturbance.

Several studies conducted in Finnmark, northern Norway and Finnish Lapland support the thesis that plant species diversity is greater in areas grazed by reindeer. Helle and Aspi (1983) and Väre et al. (1995), who concentrated their research on pine forests with a *Cladina*-dominated understory, concluded that grazing and trampling on *Cladina* have released other plants from competition with these dominant lichens. The greatest species richness found by Helle and Aspi (1983) was in moderately grazed *Calluna-Cladina*-heaths of Finnish Lapland. The species richness decreased under both lower and higher grazing pressure. Suominen and Olofsson (2000) concluded that the enriching effect of grazing on species richness is not limited to vegetation types dominated by lichens, although they found it to be strongest there. In agreement with Fox (1985), they argued that grazers can maintain the relatively high species richness of tundra plant communities by preventing any of the plant species from gaining a strong dominance.

3 Diversity Patterns in Relation to the Environment and Herbivory in Alpine Tundra Ecosystems – a Case Study

Pausas and Austin (2001) point out that it needs to be understood how patterns of species richness are related to the environment before conclusions can be drawn about the effect of biodiversity in ecosystem processes. They note diversity studies in relation to environmental gradients have been mainly single-factor studies. However, such studies are unlikely to yield the desired results, because species richness is governed by two or more environmental gradients.

In a case study (Löffler 2000, 2004, 2005, 2007, Löffler and Pape 2008), we considered these aspects. Diversity patterns in alpine tundra ecosystems of northern Norway were described and related to the environment. We hypothesized that diversity is affected by the following:

- climatic variables, such as near-surface temperature conditions and snow cover at the local scale and the oceanic-continental gradient at the regional scale (Leathwick et al. 1998, Virtanen et al. 2003, Holten 2003, Körner 2004),
- the availability of nutrients (Gould and Walker 1999, Molau 2003, Virtanen et al. 2003, Körner 2004),
- soil moisture (Leathwick et al. 1998, Körner 2004), and
- disturbance related to herbivory (Austrheim and Eriksson 2001, den Herder et al. 2003, Nagy et al. 2003, Olofsson et al. 2001, 2004, van der Wal et al. 2004).

Both α-diversity and β-diversity were studied, using vascular plants and lichens as a basis.

3.1 Study Areas

In all, 11 study areas above the alpine treeline were chosen to cover the underlying gradients of (i) the pronounced climatic differences between the sub-oceanic coastal mountains and the more continental interior areas, and (ii) differences in reindeer stocking (Fig. 1). Both acidic and non-acidic areas were studied.

3.2 Field Sampling

Vegetation surveys within each study area were conducted along transects that represented the main topographic positions and related vegetation types discriminated by snow cover and moisture (Dahl 1987). Following a stratified sampling design, each type of vegetation was represented by several samples drawn at random, each of which was one square metre in size. The percentage of coverage for a) any single plant species, b) aggregated species groups (lichens, mosses, herbs, graminoids, woody plants), and c) total vegetation were recorded. The nomenclature followed Lid and Lid (1994) for vascular plants, and Krog et al. (1994) for the lichen flora. One hundred and sixty-three species of vascular plants and lichens were recorded and used for further analyses.

One transect was chosen to be bisected by a reindeer fence, separating apparently heavily grazed areas from less grazed areas. There, eight pairs of samples were chosen to vary by grazing pressure, not environmental factors. In all, 232 samples were studied. The samples were divided into a sub-set of acidic samples (n=202) and non-acidic samples (n=30) according to soil parent material. Furthermore, for each sample, 32 environmental variables were measured or estimated indirectly (Table 1).

3.3 Data Analysis

The data analysis focused on α-diversity and β-diversity and its environmental determination. α-diversity was expressed as the number of species within each sample (species density *in sensu*, Whittaker 1975). Relationships between α-diversity and environmental variables were analysed by regression tree analysis (RTA). RTA does not involve *a priori* assumptions about any particular type of relationship between the variables, and hence is useful for capturing nonlinear relationships and facilitates the interpretation of the results (De'ath and Fabricius 2000).

Table 1 Environmental variables used for further analyses (Löffler and Pape 2008)

Altitude [m a.s.l.]	Horizon thickness [cm] of		Maximum air temperature [°C]
Topographic position (ridge, slope, ...) [nominal]	V	Dense moss layer	Minimum air temperature [°C]
Surface curvature (convex, ...) [nominal]	L	Dense litter layer	Mean air temperature [°C]
Aspect [sine and cosine transformed]	O	Dense humus layer	Annual temperature amplitude [K]
Slope [0 - 90°]	H	Peat layer	Heatsum > 5 °C [K]
Soil moisture [1 - 6, ordinal]	Ah	Humic top soil horizon	Growth period [d]
Root structure [nominal] and number in profile	Ae	Eluvial top soil horizon	Snow cover [d]
Surface skeleton cover [%]	Bh	Illuvation of humus complexes	Potential radiation during summer [kW]
pH in 5 cm depth (H_2O)	Bs	Illuvation of sesquioxides	Mean number of reindeer per km^2
pH in 15 cm depth (H_2O)	ICv	Cambic weathering horizon	Maximum number of reindeer per km^2
pH in 30 cm depth (H_2O)	S	Stagnation horizon	

The degree of differentiation among plant communities or the species turnover between samples which is referred to as β-diversity (WALKER 1995, WHITTAKER et al. 2001) was investigated using indirect ordination methods. Detrended correspondence analysis (DCA, HILL and GAUCH 1980) yielded the species turnover between samples, because ordination axes were scaled in standard deviations of species turnover. A secondary result of the DCA was the total β-diversity, expressed by the length of gradients. Furthermore, the axes were interpreted ecologically by correlation analyses to environmental variables. In addition, a direct ordination method, canonical correspondence analysis (*CCA*, TER BRAAK 1986) was applied to extract environmental variables that determine species composition and hence also differentiations among samples. Significant environmental variables were chosen by automatic forward selection routines.

3.4   α-Diversity – Effects of the Environment and Herbivory

In general, non-acidic samples showed higher species densities than their acidic counterparts ($p < 0.001$). For both, mires showed the lowest species densities with a median of nine species, whereas tall herb meadows with a median of 23 species were found to be the communities with highest species densities. Among the other vegetation types, no significant differentiation was detected.

According to RTA, species density was controlled by a combination of growing season, snow cover, pH-value, soil moisture, disturbance as indicated by thickness of the brownish lCv-horizon and bare soil cover, temperature, and precipitation, listed in order of importance. Non-acidic samples were characterized by a short growing season, but a good snow shelter showed the highest species densities, whereas very wet samples with low mean temperatures showed the lowest densities. Moist samples that have a long growing season were differentiated by disturbance, from which it may be deduced that unstable conditions corresponded to low species densities.

Regarding herbivory, samples at the heavily grazed side of the reindeer fence showed an overall decrease in species numbers as well as the percentage of coverage of vegetation compared to less grazed samples. A more differentiated view revealed a strong decline in lichen species ($p < 0.01$), lichen coverage ($p < 0.05$), and coverage of woody plants ($p < 0.05$). In contrast, graminoids were supported by heavy grazing, shown by the fact that the number of species increased significantly ($p < 0.01$) combined with a slight increase in coverage (Fig. 4).

3.5   β-Diversity: Effects of Environment and Herbivory

Long gradients of the DCA axes (axis 1 = 5.044, axis 2 = 3.814) indicated a large species turnover among samples and hence a large β-diversity. The first and most important axis (eigenvalue 0.607) was correlated with snow cover and soil moisture, as already indicated by the arrangement of vegetation type clusters along this axis. The second axis (eigenvalue 0.406) was correlated with mean temperature, length of growing season, and pH-value at 30 cm depth. Hence, local conditions (represented by the first axis) were found to be of greater importance for controlling β-diversity, whereas regional conditions (represented by the second axis) were less important. A corresponding CCA biplot of samples and environmental variables supported these findings. Samples were

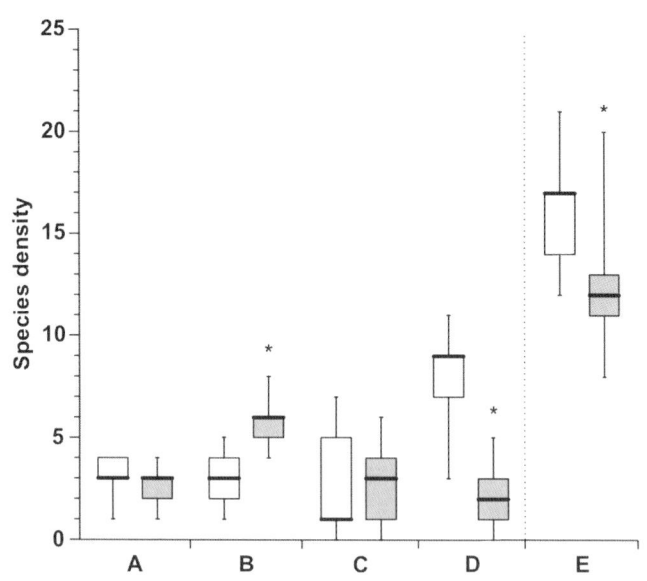

Fig. 4 Effect of heavy grazing (indicated by shaded boxes) on species richness for A woody plants, B graminoids, C herbs, D lichens, and E all. Asterisks indicate differences that were significant at the 0.01-level (modified after Löffler and Pape 2008).

arranged according to local moisture and snow cover gradients as well as regional mean temperatures, temperature amplitude, and growing season gradients. In contrast to the specific case at the reindeer fence, herbivory did not contribute significantly to the differentiation among samples for the entire set of samples.

4 Plant Diversity Patterns and Reindeer Pastoralism in Northern Norwegian Mountain Systems: Status Quo and Future Challenges

Given that the summer and winter diets of reindeer differ markedly, the impacts of reindeer grazing on winter and summer pastures could be expected to be different. Generally, all northern ecosystems have been shaped by grazing and trampling since the last glacial epoch. It is evident that grazing affected both community composition and the density of specific taxonomical groups. However, as reindeer husbandry developed very early, it is difficult to quantify the impact of human-controlled reindeer pastoralism on a "pristine" landscape. Nevertheless, the technical revolution in reindeer husbandry, starting with the introduction of snow scooters in the 1960s, must have had strong effects on the ecosystems. The number of reindeers has increased since than by about 400 percent as a consequence of the change from pastoralism to the profit-oriented production of meat. Moreover, the availability of pastures decreased at the same time, due to loss of area and disturbance effects related to the extension of infrastructure, power supply facilities, houses, and cabins.

All the studies reviewed emphasized the importance of grazing pressure when regarding the effects of grazing on ecosystems. Inevitably, the issue arises whether actual stocking sizes lie below or above the certain, but yet unknown, threshold at which a positive feedback of grazing turns into a negative one. To make things worse, such a threshold need not to be the same for different ecosystem properties (e. g. primary production, diversity) and ecosystems (summer/winter pastures).

Studies of OLOFSSON et al. (2001, 2004) on summer pastures have shown that artificially high grazing pressure in the vicinity of fences was able to create steppe-like vegetation dominated by graminoids. With the transition from heathland to grassland, both primary production and food quality increased, because graminoids are preferred to ericoid dwarf shrubs as a source of food (BATZLI et al. 1980). Following ZIMOV et al. (1995), such a steppe stage of vegetation supporting high densities of herbivores, which was also common in cold regions until the end of the Pleistocene, could be an alternative steady state for other types of tundra even under current climatic conditions. The results of OLOFSSON's et al. (2001, 2004) studies also suggested that moderate grazing might not be the best way to use the summer ranges. Under such conditions, the reindeer fed selectively, favouring the abundance of non-preferred plants. At higher grazing pressure, the herbivores tended to be less selective and the importance of trampling increased. Thus, preferred food plants with high grazing tolerance, such as graminoids, may be favoured by intensive grazing (WESTOBY et al. 1989). On the basis of these findings, OLOFSSON et al. (2001, 2004) suggested that when creating productive summer grazing ranges, the focus should be on keeping the vegetation in the favourable state where the preferred food source (graminoids) dominates the vegetation and primary production is high. This might be best achieved by periodic intense grazing interspersed with periods of low grazing pressure. However, more recently, BRÅTHEN et al. (2007) warned explicitly against such a management strategy on the basis of the results of their own large-scale study in the same area, which represented about 62 percent of the summer ranges in Finnmark. They found that forage resources were depressed where reindeer occurred at high densities.

Several studies, including our own case study, provided evidence that the herbivore-pasture system in northern Norway is, with its present stocking rates, in a transitional stage towards steppe-like vegetation. For summer pastures, appears beneficial from an economic point of view as it results in enhanced productivity sustaining reindeer pastoralism. However, these positive effects could be scrutinized for winter pastures. A dramatic degradation of lichen pastures has been reported by numerous studies (e. g. EVANS 1996, JOHANSEN and KARLSEN 1998) and explained as a consequence of the excessive pressure imposed on the resource by stocking densities that are too great. Here, the process that is beneficial for the productivity of summer pastures might reverse, because lichens are generally supposed to constitute the main winter fodder of reindeer. Although vegetation shifts were shown to occur in a variety of heathlands, from dry and nutrient-poor to moist and nutrient-rich (OLOFSSON et al. 2001), they did not all respond in the same way to grazing. On dry exposed ridges, graminoids did not seem to increase when dwarf shrubs, mosses, and lichens are reduced and, as a consequence, in these habitats grazing led to areas largely devoid of vegetation (EVANS 1996). Yet these are the habitats important for winter pastures! However, studies on the composition of the winter diet (GAARE and SKOGLAND 1975, ERIKSSON 1981; STAALAND and SÆBØ 1987; MATHIESEN et al. 2000) have shown that reindeer select a mixed diet in winter as well as in summer, with lichens constituting about 37-73 percent, followed by woody plants (13-45 %), and graminoids with 8-29 percent (Storeheier et al. 2003). Consequently, van der Wal (2006) concluded that although reindeer feed on lichens, they do not depend critically on them and they may switch over to other plants. Hence, the loss of lichens on winter pastures (provided that other palatable plants are available) might not be as dramatic for reindeer pastoralism as is suggested in the literature.

However, turning from economic to ecological aspects (e. g. diversity), the current development of pastures might be seen in another light. Despite the fact that diversity was found to be favoured by a distinct grazing pressure that prevents any plant species from outcompeting the others, as well as shaping different successional stages and micro-habitats due to disturbance, we expect diversity to suffer from the transition towards steppe-like vegetation with dominating graminoids. Olofsson (2006) found contrasting results for inherent species-rich and species-poor communities. Species-rich plant communities were less resilient to grazing with a corresponding decline in species richness, while species richness increased in inherent species-poor communities, corresponding to a homogenization of the landscape. A decrease in species richness due to heavy grazing (the most pronounced effects being on bryophytes and lichens) is also reported by Eskelinen and Oksanen (2006). In general, the effect of herbivory on the maintenance of plant diversity in alpine environments is inadequately understood and needs further research (Erschbamer et al. 2003). Our case study revealed an overall reduction of diversity due to heavy grazing at nutrient-poor sites. In accordance with the theory of an herbivore-driven shift in vegetation, we found only graminoids to be supported by grazing pressure, whereas lichens suffered the most (Löffler 2004, 2007).

In addition to the herbivore-pasture system being affected by the indigenous factors that are discussed above, it is also subject to exogenous influences. Habitat loss due to increasing competition for the available land and resulting higher grazing pressure on the remaining pastures is likely to enhance the outlined processes. Furthermore, climatic variation has been reported to affect several life-history and population parameters of northern ungulates (Gaillard et al. 2000; Ottersen et al. 2001; Weladji et al. 2002). The effect might be direct (because during winter increasing snow depth leads to increasing costs for locomotion while forage availability decreases), or indirect, (because during summer abundance, the quality and production of forage plants is affected) (Weladji and Holand 2003).

The highly emotive debate about reindeer management, which is fuelled by terms such as "environmental catastrophe" or "ecological disaster", has resulted in reindeer pastoralism being viewed, in some quarters, as synonymous with habitat degradation. However, van der Wal (2006) used the concept of alternative stable states to evaluate herbivore-driven vegetation change and to assist in the differentiation between vegetation state transition and habitat degradation. He thereby defused a central environmental debate. He concluded that an acknowledgement of the existence of alternative vegetation states in tundra ecosystems might be useful for evaluating the observed radical changes in vegetation that are occurring throughout the reindeer range. The relatively rapid changes in vegetation thus do not necessarily equate to habitat degradation, but in many cases reflect predictable changes in the vegetation of mountain systems.

5   References

AUGUSTINE, D. J.; L. E. FRELICH; P. A. JORDAN (1998): Evidence for two alternate stable states in an ungulate grazing system. Ecological Applications 8: 1260–1269.

AUSTRHEIM, G.; O. ERIKSSON (2001): Plant species diversity and grazing in the Scandinavian mountains – patterns and processes at different spatial scales. Ecography 24: 683–695.

BATZLI, G. O.; R. G. WHITE; S. F. MACLEAN; F. A. PITELKA; B. D. COLLIER (1980): The herbivore-based trophic system. In: BROWN, J.; R. G. MILLER; L. L. TIESZEN; L. L. BUNNELL (eds.): An arctic ecosystem. The coastal tundra at Barrow, Alaska. US/IBP Synthesis Series. PA: Dowden, Hutchinson and Ross. pp. 335–410.

BAZELY, D. R.; R. L. JEFFERIES (1986): Changes in the composition and standing crop of salt-marsh communities in response to the removal of a grazer. Journal of Ecology 74: 693–706.

BENISTON, M. (2003): Climatic change in mountain regions. A review of possible impacts. Climatic Change 59: 5–31.

BOUDREAU, S.; S. PAYETTE (2004): Caribou-induced changes in species dominance of lichen woodlands: An analysis of plant remains. American Journal of Botany 91(3): 422–429.

BRÅTHEN, K. A.; J. OKSANEN (2001): Reindeer reduce biomass of preferred plant species. Journal of Vegetation Science 12: 473–480.

BRÅTHEN, K. A.; R. A. IMS; N. G. YOCCOZ; P. FAUCHALD; T. TVERAA; V. H. HAUSNER (2007): Induced Shift in Ecosystem Productivity? Extensive Scale Effects of Abundant Large Herbivores. Ecosystems 10: 773–789.

BRONNY, H.; I. HEMMER; N. T. SOKKI (1985): Samische Rentierwirtschaft – Reliktform oder Wachstumsbranche. Geographische Rundschau 37(10): 529–536.

BURKHARD, B. (2004): Ecological Assessment of the Reindeer Husbandry System in Northern Finland. EcoSys, Suppl. Band 43.

CRAWLEY, M. J. (1997): Plant-herbivore dynamics. In: CRAWLEY, M. J. (ed.), Plant ecology. Blackwell: pp. 401–474.

CRÊTE, M.; G. J. DOUCET (1998): Persistent suppression in dwarf birch after release from heavy summer browsing by caribou. Arctic and Alpine Research 30: 126–132.

DANELL, K.; P. M. UTSI, R. T. PALO, O. ERIKSSON (1994): Food plant selection by reindeer during winter in relation to plant quality. Ecography 17: 153–158.

DE'ATH, G.; K. E. FABRICIUS (2000): Classification and regression trees: a powerful yet simple technique for the analysis of complex ecological data. Ecology 81: 3178–3192.

DEN HERDER, M.; M.-M. KYTÖVIITA; P. NIEMELÄ (2003): Growth of reindeer lichens and effects of reindeer grazing on ground cover vegetation in a Scots pine forest and a subarctic heathland in Finnish Lapland. Ecography 26: 3–12.

DEN HERDER, M.; VIRTANEN, R. and H. ROININEN (2004): Effects of reindeer browsing on tundra willow and its associated insect herbivores. Journal of Applied Ecology 41: 870–879.

DIAZ, H. F.; M. GROSJEAN; L. GRAUMLICH (2003): Climate variability and change in high elevation regions: Past, present and future. Climatic Change 59: 1–4.

ERIKSSON, O. (1981): Renens vinterdiet. Växtekologiska studier 13: 25–46.

ERSCHBAMER, B.; R. VIRTANEN; L. NAGY (2003): The impact of vertebrate grazers on vegetation in European high mountains. In: NAGY, L.; G. GRABHERR; C. KÖRNER; D. B.

A. Thompson (eds.), Alpine biodiversity in Europe. Ecological Studies 167. Berlin, Springer, pp. 379–396.

Eskelinen, A.; J. Oksanen (2006): Changes in the abundance, composition and species richness of mountain vegetation in relation to summer grazing by reindeer. Journal of Vegetation Science 17: 245–254.

Evans, R. (1996): Some impacts of overgrazing by reindeer in Finnmark, Norway. Rangifer 16: 3–19.

Fosaa, A. M. (2004): Biodiversity patterns of vascular plant species in mountain vegetation in the Faroe Islands. Diversity and Distributions 10: 217–223.

Fox, J. F. (1985): Plant diversity in relation to plant production and disturbance by voles in Alaskan tundra communities. Arctic and Alpine Research 17: 199–204.

Gaare, E.; T. Skogland (1975): Wild reindeer food habits and range use at Hardangervidda. In: Wielgolaski, F. E. (ed.): Fennoscandian Tundra Ecosystems. Part 2: Animals and Systems Analysis, pp. 195–205. Berlin: Springer.

Gaillard, J.-M.; M. Festa-Bianchet; N. G. Yoccoz; A. Loison; C. Toigo (2000): Temporal variation in fitness components and population dynamics of large herbivores. Annu Rev Ecol Syst 31: 367–393.

Gough, L.; G. R. Shaver; J. Carroll; D. L. Royer; J. A. Laundre (2000): Vascular plant species richness in Alaskan arctic tundra: the importance of soil pH. Journal of Ecology 88: 54–66.

Gould, W. A.; M. D. Walker (1999): Plant communities and landscape diversity along a Canadian Arctic river. Journal of Vegetation Science 10: 537–548.

Grellmann, D. (2002): Plant responses to fertilization and exclusion of grazers in an arctic tundra heath. Oikos 98: 190–204.

Grytnes, J. A. (2003): Species-richness patterns of vascular plants along seven altitudinal transects in Norway. Ecography 26: 291–300.

Heikkinen, R. K.; Neuvonen (1997): Species richness of vascular plants in the subarctic landscape of northern Finland: modelling relationships to the environment. Biodiversity and Conservation 6: 1181–1201.

Helle, T.; J. Aspi (1983): Effects of winter grazing by reindeer on vegetation. Oikos 40: 337–343.

Hill, M. O.; H. G. Gauch (1980): Detrended correspondence analysis: an improved ordination technique. Vegetatio 42: 47–58.

Hobbs, R. J.; L. F. Huenneke (1992): Disturbance, diversity and invasion: implications for conservation. Conservation Biology 6: 324–337.

Holten, J. I. (2003): Altitude ranges and spatial patterns of alpine plants in northern Europe. In: Nagy, L.; G. Grabherr; C. Körner; D. B. A. Thompson (eds.), Alpine biodiversity in Europe. Ecological Studies 167. Berlin, Springer: pp. 173–184.

Huston, M. A. (1979): A general hypothesis of species diversity. American Naturalist 113: 81–101.

Johansen, B.; S. R. Karlsen (1998): Endringer i lavdekket på Finnmarksvidda 1987–96, basert på Landsat 5-TM data. Tromsø: NORUT.

Kalela, O.; T. Koponen; E. A. Lind et al. (1961) Seasonal change of habitat in the Norwegian lemming, Lemmus lemmus (L.). Ann. Acad. Sci. Fenn. Ser. A IV Biol. 55: 1–72.

Körner, C. (1995): Alpine plant diversity: A global survey and functional interpretations. In: Chapin III, F. S.; C. Körner (eds.), Arctic and alpine biodiversity. Ecological Studies 113. Berlin, Springer: pp. 45–62.

Körner, C. (2003): Alpine plant life: functional plant ecology of high mountain ecosystems. Berlin, Springer.
Körner, C. (2004): Mountain biodiversity, its causes and function. Ambio, Special Report 13: 11–17.
Krog, H., H. Østhagen; T. Tønsberg (1994): Lavflora. Norske busk- og bladlav. Oslo, Universitetsforlaget.
Leathwick, J. R.; B. R. Burns; B. D. Clarkson (1998): Environmental correlates of tree alpha-diversity in New Zealand primary forests. Ecography 21: 235–246.
Lid, J.; D. T. Lid (1994): Norsk flora. Oslo, Det Norske Samlaget.
Löffler, J. (2000): High mountain ecosystems and landscape degradation in northern Norway. Mountain Research and Development 20: 356–363.
Löffler, J. (2004): Degradation of high mountain ecosystems in northern Europe. Journal of Mountain Science 2: 97–115.
Löffler, J. (2005): Reindeer grazing impact on arctic-alpine landscapes in western Greenland, and central and northern Norway. Geoöko 26: 1–18.
Löffler, J. (2007): Reindeer grazing changes diversity patterns in arctic-alpine landscapes of northern Norway. Die Erde 138: 215–233.
Löffler, J.; R. Pape (2008): Diversity Patterns in Relation to the Environment in Alpine Tundra Ecosystems of Northern Norway. Arctic, Antarctic, and Alpine Research 40: 373–382.
Lomolino, M. V. (2001): Elevation gradients of species-diversity: historical and prospective views. Global Ecology and Biogeography 10: 9–13.
Manseau, M.; J. Huot; M. Crête (1996): Effects of summer grazing by caribou on composition and productivity of vegetation: community and landscape level. Journal of Ecology 84: 503–513.
Mathiesen, S. D.; Ø. E. Haga; T. Kaino; N. J. C. Tyler (2000): Diet composition, rumen papillation and maintenance of carcass mass in female Norwegian reindeer (*Rangifer tarandus tarandus*) in winter. Journal of Zoology 251: 129–138.
McNaughton, S. (1984): Grazing lawns: animals in herds, plant form and coevolution. American Naturalist 124: 863–886.
Messerli, B.; J. D. Ives (1997): Mountains of the world: A global priority. New York, Parthenon.
Milchunas, D. G.; O. E. Sala; W. K. Lauenroth (1988): A generalized model of effect of grazing by large herbivores in grassland community structure. American Naturalist 132: 87–106.
Molau, U. (2003): Overview: Patterns in diversity. In: Nagy, L.; G. Grabherr; C. Körner and D. B. A. Thompson (eds.), Alpine biodiversity in Europe. Ecological Studies 167. Berlin, Springer: pp. 125–132.
Moser, D.; S. Dullinger; T. Englisch; H. Niklfeld; C. Plutzar; N. Sauberer; H. G. Zechmeister; G. Grabherr (2005): Environmental determinants of vascular plant species richness in the Austrian Alps. Journal of Biogeography 32: 1117–1127.
Mulder, C. P. H.; R. W. Ruess (1998): Effects of herbivory on arrowgrass: interactions between geese, neighbouring plants, and abiotic factors. Ecological Monographs 68: 275–293.
Nagy, L.; G. Grabherr; C. Körner; D. B. A. Thompson (2003): Alpine biodiversity in space and time: a synthesis. In: Nagy; L., G. Grabherr; C. Körner; D. B. A. Thompson (eds.), Alpine biodiversity in Europe. Ecological Studies 167. Berlin, Springer: pp. 453–464.

Oksanen, L.; T. Oksanen (2000): The logic and realism of the hypothesis of exploitation ecosystems. American Naturalist 155: 703–723.

Oksanen, L.; S. D. Fretwell; J. Arruda; P. Niemelä (1981): Exploitation ecosystems in gradients of primary productivity. American Naturalist 118: 240–261.

Olff, H.; M. E. Ritchie (1998): Effects of herbivores on grassland plant diversity. Trends Ecol. Evol. 13: 261–265.

Olofsson, J. (2006): Plant Diversity and Resilience to Reindeer Grazing. Arctic, Antarctic, and Alpine Research 38: 131–135.

Olofsson, J.; H. Kitti; P. Rautiainen; S. Stark; L. Oksanen (2001): Effects of summer grazing by reindeer on composition of vegetation, productivity and nitrogen cycling. Ecography 24: 13–24.

Olofsson, J.; J. Moen; L. Oksanen (2002): Effects of herbivory on competition intensity in two arctic-alpine tundra communities with different productivity. Oikos 96: 265–272.

Olofsson, J.; S. Stark; L. Oksanen (2004): Reindeer influence on ecosystem processes in the tundra. Oikos 105: 386–396.

Ottersen, G.; B. Planque; A. Belgrano; E. Post; P. C. Reid; N. C. Stenseth (2001): Ecological effects of the North Atlantic Oscillation. Oecologia 128: 1–14.

Pausas, J. G.; M. P. Austin (2001): Patterns of plant species richness in relation to different environments: An appraisal. Journal of Vegetation Science 12: 153–166.

Proulx, M.; A. Mazumder (1998): Reversal of grazing impact on plant species richness in nutrient-poor vs. nutrient-rich ecosystems. Ecology 79: 2581–2592.

Reindriftsforvaltningen (2007): Ressursregnskap for reindriftsnæringen. For reindriftsåret 1. April 2005 – 31. Mars 2006. http://www.reindrift.no (20.01.2008).

Riseth, J. Å.; B. Johansen; A. Vatn (2004): Aspects of a two-pasture – herbivore model. Rangifer 15: 65–81.

Rosenzweig, M. L. (1995): Species diversity in space and time. Cambridge, Cambridge Univ. Press.

Sánchez-Gonzáles, A.; L. López-Mata (2005): Plant species richness and diversity along an altitudinal gradient in the Sierra Nevada, Mexico. Diversity and Distributions 11: 567–575.

Staaland, H.; S. Sæbø (1987): Seasonal variations in mineral status of reindeer calves from Elgaa reindeer herding district, Norway. Rangifer 7: 22–28.

Stark, S.; D. Grellmann (2002): Soil microbial responses to mammalian herbivory in an arctic tundra heath at two levels of nutrient availability. Ecology 83: 2736–2744.

Storeheier, P. V.; B. E. H. van Oort; M. A. Sundset; S. D. Mathiesen (2003): Food intake of reindeer in winter. Journal of Agricultural Science 141: 93–101.

Suominen, O.; J. Olofsson (2000): Impacts of semi-domesticated reindeer on structure of tundra and forest communities in Fennoscandia: A review. Ann. Zool. Fennici 37: 233–249.

ter Braak, C. J. F. (1986): Canonical correspondence analysis: a new eigenvector technique for multivariate direct gradient analysis. Ecology 67: 1167–1179.

Thompson, L. G. (2000): Ice core evidence for climate changes in the tropics: Implications for our future. Quat. Sci. Rev. 19: 19–35.

van der Wal, R. (2006): Do herbivores cause habitat degradation or vegetation state transition? Evidence from the tundra. Oikos 104: 177–186.

VAN DER WAL, R.; R. D. BARDGETT; K. A. HARRISON; A. STIEN (2004): Vertebrate herbivores and ecosystem control: cascading effects of faeces on tundra ecosystems. Ecography 27: 242–252.

VÄRE, H., R. OHTONEN; J. OKSANEN (1995): Effects of reindeer grazing on understorey vegetation in dry *Pinus sylvestris* forests. J. Veg. Sci. 6: 523–530.

VIRTANEN, R. (1998): Impact of grazing and neighbour removal on a heath plant community transplanted onto a snowbed site, NW Finnish Lapland. Oikos 81: 359–367.

VIRTANEN, R. (2000): Effects of grazing on above-ground biomass on a mountain snowbed, NW Finland. Oikos 90: 295–300.

VIRTANEN, R., H. HENTTONEN; K. LAINE (1997): Lemming grazing and structure of a snowbed plant community – a long term experiment at Kilpisjärvi, Finnish Lapland. Oikos 79: 155–166.

VIRTANEN, R., T. DIRNBÖCK; S. DULLINGER; G. GRABHERR; H. PAULI; M. STAUDINGER; L. VILLAR (2003): Patterns in plant species richness of European high mountain vegetation. In: NAGY, L.; G. GRABHERR; C. KÖRNER; D. B. A. THOMPSON (eds.), Alpine biodiversity in Europe. Ecological Studies 167. Berlin, Springer: pp. 149–172.

WALKER, M. D. (1995): Patterns and causes of Arctic plant community diversity. In: CHAPIN, F. S.; C. KÖRNER (eds.), Arctic and alpine biodiversity. Ecological Studies 113, Berlin, Springer: pp. 3–20.

WEGENER, C.; A. ODASZ-ALBRIGTSEN (1998): Do Svalbard reindeer regulate standing crop in the absence of predators? A test of the 'exploitation ecosystems' model. Oecologia 116: 202–206.

WELADJI, R. B.; D. R. KLEIN; Ø. HOLAND; A. MYSTERUD (2002): Comparative response of *Rangifer tarandus* and other northern ungulates to climatic variability. Rangifer 22: 33–50.

WELADJI, R. B.; Ø. HOLAND (2003): Global climate change and reindeer: effects of winter weather on the autumn weight and growth of calves. Oecologia 136: 317–323.

WESTOBY, M.; B. WALKER; I. NOY-MEIR (1989): Opportunistic management for rangelands not at equilibrium. Journal of Range Management 42: 266–274.

WHITTAKER, R. H. (1975): Communities and Ecosystems. Houndmills, Macmillan.

WHITTAKER, R. J.; K. J. WILLIS; R. FIELD (2001): Scale and species richness: towards a general, hierarchical theory of species diversity. Journal of Biogeography 28: 453–470.

ZIMOV, S. A.; V. I. CHUPRYNIN; A. P. ORESHKO; F. S. CHAPIN III; J. F. REYNOLDS; M. C. CHAPIN (1995): Steppe-tundra transition: a herbivore driven biome shift at the end of the pleistocene. Am. Nat. 146: 765–794.

Agrarian Diversity, Resilience and Adaptation of Andean Agriculture and Rural Communities

Christoph Stadel[1]

> "The most profound meaning of the Andes thus comes not from a physical description, but from the cultural outcome of 10 millennia of knowing, using, and transforming the varied environments of western South America." Gade 1999: 34

Introductory Remarks

Tropical mountains, in many parts, offer ecological conditions for agriculture that are superior to those of adjacent lowlands. These environmental assets and the potential agricultural wealth have significantly contributed to a sustained human settlement and to the rise of major civilizations. Tropical mountains also exhibit an intriguing diversity of agricultural landscapes, farming systems and rural livelihoods reflecting not only highly varied natural environments, but also distinct cultural traditions, social conditions, economic activities, and political ecologies. This horizontal and altitudinal zonation of ecological and agricultural regions, of distinct watersheds and niches has been well documented in the literature. While tropical mountains may offer a good potential for agricultural activities, they are also environments with an array of threatening natural stressors, risks and hazards. Furthermore, being largely located within a realm of prevailing economic and social underdevelopment, farming communities have often to face conditions of poverty marginalization (Stadel 1991).

In order to cope with the criticality and vulnerability of the natural and human environments, rural populations have resorted to various forms of resilience and adaptation (Knapp 1991). Resilience, in a wider sense, may be defined as the ability of systems and communities to cope with adverse conditions, to absorb or buffer negative impacts of changes, to adapt to new situations, or even to rebound and gain new strength (Resilience Alliance, no date, Bohle 2008, Holling 1973, Janssen & Orstrom 2006). Farmers in tropical mountains and in different cultural contexts have given numerous testimonies of resilience and adaptations, therefore it appears paramount to promote or mobilize this human potential in any rural development initiatives.

Agrarian Diversity in the Tropical Andes

The tropical Andes count among the mountain landscapes with the greatest agrarian diversity (Brush 1987). This agrarian diversity of the Andes reflects a corollary of highly varied locational, ecological, cultural, social, economic and political conditions (refer to Figs. 1 and 2):

- regional variations: in terms of geomorphology, soils, climate, vegetation cover and human environments,
- altitudinal variations of ecological conditions, cultigens and agricultural systems,

---

[1] University of Salzburg, Department of Geography, A-5020 Salzburg, christoph.stadel@sbg.ac.at.

- a great variety of local or regional ecological and agricultural niches,
- impact of degradation, as well as of conservation efforts,
- influence of hydrographic conditions, water tenure, water supply and irrigation methods,
- accessibility to roads and markets, degree of market penetration,
- impact of ethnic and cultural groups, influence of demographic patterns,
- adherence to heritage and tradition, or adoption of innovations, new technologies or modernization,
- impact of exogenous actors, policies or programs.

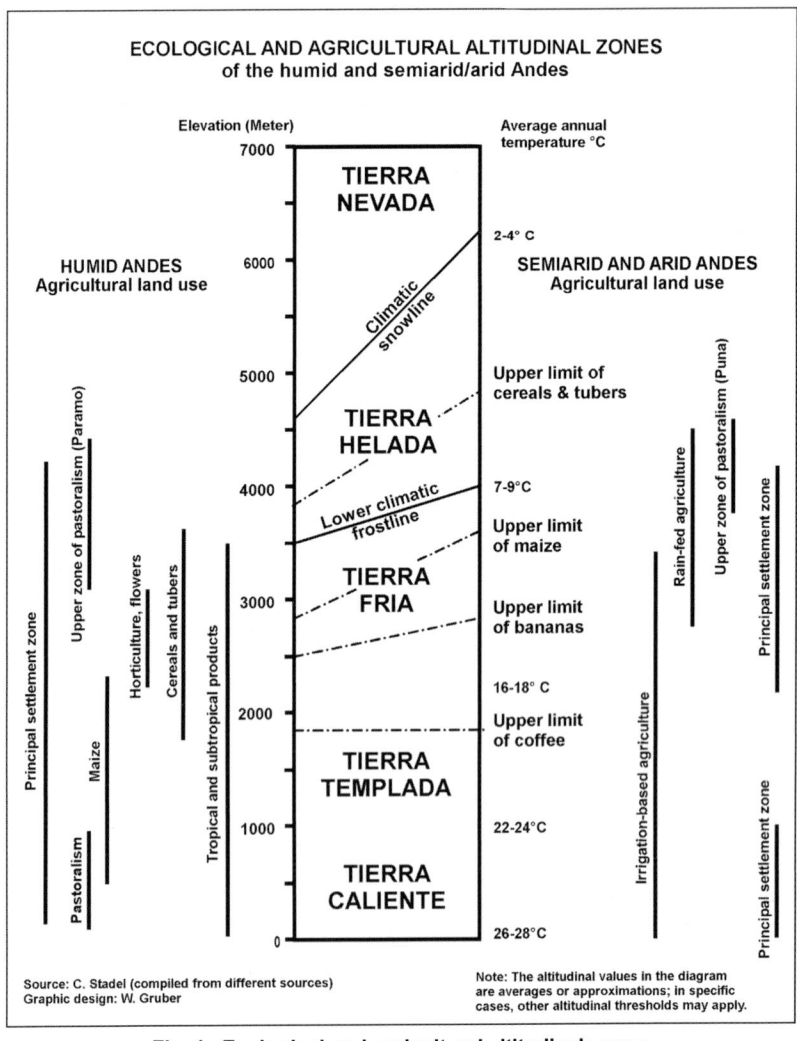

Fig. 1 Ecological and agricultural altitudinal zones

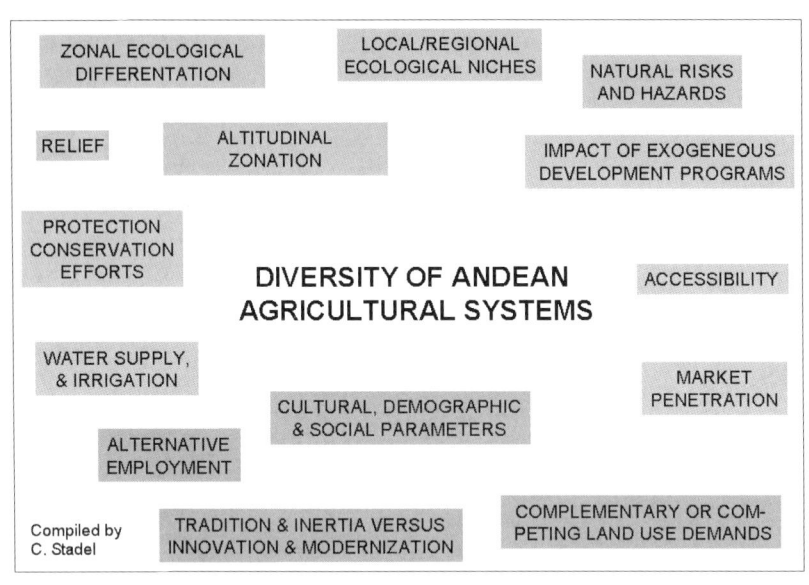

Fig. 2 Diversity of Andean agricultural systems

Table 1 summarizes the major assets and constraints of Andean agriculture. The variety of these assets or positive stimuli and stress factors suggest that the agricultural landscape of the Andes contains both niches of opportunities, as well as vulnerable spaces.

ZIMMERER (1999) has characterized the diversity of Andean agricultural landscapes, as "overlapping patchworks of mountain agriculture". He emphasized not only their spatial variety, but also the dynamic nature and temporal changes of farming systems. BEBBINGTON (1997) underlines the importance of those "islands of sustainability" of mountain agriculture and rural livelihoods which can serve as diffusing promoters for a sustainable regional development!

In the spatial system of *Mitimagkuna* or "Vertical Control of different ecological zones", Andean communities utilized the potential and resources of such *archiepélagos verticales*. At each altitudinal zone or niche, specific crops were cultivated and were often combined with various forms of animal husbands, and pastoralism. Also, multiple forms of economic exchange were carried out between highlands and lowlands and between different altitudinal zones.

In the 1980s, the vertical patterns of difference "as a general underpinning of human life in the mountain" (HEWITT 1988: 22) and the overall applicability of contentional altitudinal zonation models of mountain land use were questioned. ALLAN (1986) proposed instead an "accessibility model" of mountain land use for situations where highways and roads are penetrating traditional mountain landscapes:

> *"Accessibility provides avenues for the diffusion of ideas, technology, and goods into the mountains [...]. Roads have had a profound impact upon the area in close proximity to them, and in mountain regions this has led to a change in land use."* (ALLAN 1986: 186)

Table 1 Assets and Constraints

ASSETS OF ANDEAN AGRICULTURE
- FAVORABLE CLIMATIC AND SOIL CONDITIONS
- GOOD WATER SUPPLY
- FERTILE SOILS
- LOCAL EXPERIENCE AND EXPERTISE OF FARMERS AND THEIR ATTACHMENT TO THE SOIL
- GENETIC POOL OF SEED AND BREED VARIETIES
- DIVERSITY AND COMPLEMENTARITY OF AGRICULTURAL PRODUCTS
- GOOD POTENTIAL OF MANY PRODUCTS FOR REGIONAL; NATIONAL AND
- INTERNATIONAL MARKETS

CONSTRAINTS FOR ANDEAN AGRICULTURE
- MULTIPLE NATURAL RISKS AND HAZARDS (EARTHQUAKES, VOLCANISM, ARIDITY, FLOODING, FROST, PESTS, EROSION AND LAND DEGRADATION, CONTAMINATION)
- ADVERSE RELIEF CONDITIONS; ALTITUDINAL THRESHOLDS, ACCESSIBILITY BARRIERS
- POOR OR DEGRADED SOILS
- RURAL POPULATION PRESSURE, BUT ALSO DEGRADATION OF AGRICULTURE AND ABANDONMENT OF FIELDS RESULTING FROM OUT-MIGRATION
- ADVERSE AND UNJUST LAND TENURE AND WATER DISTRIBUTION SYSTEM
- LIMITED MARKET ACCESS, INADEQUATE TRANSPORTATION SYSTEM
- INADEQUATE CAPACITATION AND FORMATION OF FARMERS, INSUFFICIENT TECHNICAL SUPPORT
- INADEQUATE RURAL FINANCIAL GRANTS AND CREDITS FOR AGRICULTURAL DEVELOPMENT
- ATTITUDES OF INERTIA, PASSIVITY AND PESSIMISM WITHIN FARMING COMMUNITIES
- NEGATIVE IMPACTS OF OUSIDE INTERFERENCES

Compiled by C. Stadel

While a number of authors welcomed Allan's contribution as a "challenge to the time honoured bio-physical, or geoecological, approach to mountain research" (IVES 1986: 183) others voiced criticism for Allan's sweeping statement that "the altitudinal zonation model is no longer suitable for characterizing mountain ecosystems now that human activity is directed to new motorized transportation networks limited to a wider political economy and no longer dependent on altitude" (ALLAN 1986: 185). In making reference to the rich variety of "horizontal", "vertical" and "aspectual" landscapes of Ecuador, STADEL (1992) lists – in addition to the ecological conditions – the following factors which had an impact on the agricultural land use and on rural livelihoods in the Andean mountains:

- The cultural environment, in particular the contrast between the regions with a predominant White/Mestizo population and the settlement areas of highland and lowland Indian groups, with their influence on land tenure, agricultural practices and settlement forms.
- The age and the nature of the settlement process with its inherent imprint on the cultural landscape.
- The relative distance to and accessibility of market centers and highways which tend to influence population densities, employment and mobility patterns, land prices and land use.
- The provision of irrigation water, traditional water rights and new irrigation schemes which led to significant contrasts in the intensity of land use and in the types of crops grown.
- The nature of economic opportunities and constraints within the region itself and in neighboring regions.
- Local and regional market conditions and the influence of local elites on economic activities.
- The influence of local traditions, local organization, perceptions and initiatives.
- External influences by governmental and non-governmental, regional, national or international agencies and organizations. This may lead to noticeable changes in agricultural practices, land use, infrastructures and services and settlement patterns.

Vulnerability, Resilience and Adaptations of Andean Agriculture

A principal objective of Andean agriculture is to cope with ecological, economic and social crisis situations which result in various manifestations of vulnerability. Examples for ecological crises and vulnerability include a resource scarcity, a deterioration of resource quality, an unsustainable exploitation of the environment, or inadequate conservation efforts. Social crises and vulnerability may result from inadequate social security and a collapse of social networks. Economic crises and vulnerability may emanate from impoverishment, insufficient income, food, and in general marginal livelihoods (Fig. 3). Fig. 3 also portrays potential survival strategies of Andean communities in their attempt to cope with a critical situation. Ecological strategies may include resource protection and conservation efforts or the recurrence to alternative forms of resource use, with the objective of enhancing the environmental quality in a sustained fashion. Social survival strategies seek to improve the forms of social and organization and networking, with the goal of strengthening the local social political power. Economic survival strategies aim at improving the economic situations of families and village communities, for instance by securing an adequate subsistence and market-oriented agricultural production, or by resorting to alternative employment opportunities and income.

The autochthonous population, for a long time, has adapted to the physiographic and ecological diversity of the Andes by developing highly complex agro-pastoral systems and livelihoods. These were rooted in a traditional Andean culture which Gade (1999: 36) summarized under the term of *lo Andino:*

ECONOMIC	SOCIAL	ECOLOGICIAL
Empoverishment Marginalization	**CRISIS SITUATION** Collapse of social Networks	Ressource scarcity, deterioration of resource quality
Marginal livelihoods, insufficient income, food	**VULNERABILITY** Inadequate social security	Inadequate conservation, non-sustainable exploitation of environment
Enhancement of subsistence production alternative employment and incomes	**SURVIVAL STRATEGIES** Improved forms of organization and networking	Conservation, resource protection alternative resource use
Improvement of economic situation	Enhancement of local social and political power	Enhancement of environmental quality

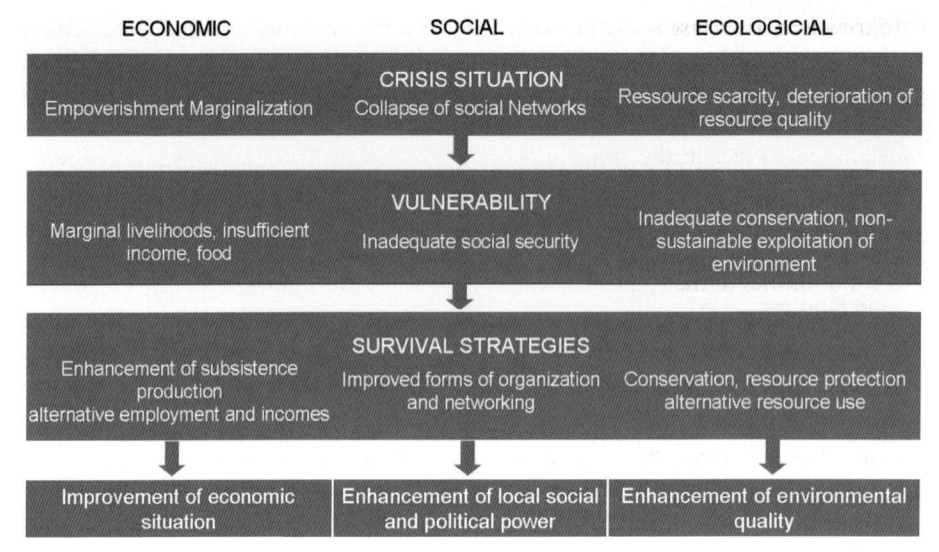

Fig. 3   **Vulnerability & survival strategies**

"Many autochthonous elements, practices, strategies and symbols both material and non-material, make up the sum of lo Andino."

The objective of the agricultural pursuits of Andean communities was to minimize the environmental and economic risks and vulnerabilities by an optimal utilization of the potential of the topography, climate, soil and hydrographic conditions for field cultivation and pastoralism. A variety of cultigens originated in the tropical Andes and were later widely diffused, others remained largely confined to the Andean realm and continued to be cultivated by the autochtonous population. This was complemented by a domestication of animals, e. g. llamas, alpacas, vicuñas, guinea pigs, as well as by various forms of preparing and conserving agricultural products, for instance the preparation of *chuño* produced from special types of high-altitude tuber crops by a process of repeated freezing and drying (STADEL 2001).

Minimizing the environmental risks and enhancing the natural potential of the land and water resources was also the determining objective of terracing mountain slopes, of various irrigation systems, of intercropping practices, of different forms of crop and field rotation cycles, and of the use of specific cultivation tools and techniques (VOGL 1990). Furthermore, the promotion of an "agro-biodiversity", for instance by organizing seed banks and fairs (TAPIA 2000), or of an "agro-ecology" (DELGADO 1993) and "Bio-Cultural Diversity" are strategies for enhancing an endogeneous agricultural development (RIST 2007). While many of these strategies and methods are rooted in a proven traditional *Saber Andino*, it should be underlined that the activities of Andean farmers have never been immutable but have frequently been adjusted to newly occurring risks or to new opportunities, e. g. the cultivation of products with a strong "market appeal", or the adoption of new irrigation and breeding techniques and plant or animal disease control measures.

Rhoades and Thompson (1975: 547) suggested two types of adaptive strategies of Andean communities: a "generalized type" in which a single population "through agro-pastoral transhumance directly exploits a series of micro-niches or eco-zones at several altitudinal levels", and a "specialized type" in which a population "locks into a single zone and specializes in the agricultural or pastoral activities suitable to that altitude, developing elaborate trade relationships with populations in other zones which are also involved in specialized production". Thomas, Winterhalder and McRae, (1979) making reference to the multiple resource base and the behavioral flexibility of individuals and communities of the tropical Andes, distinguished the following five major agricultural "response types": rotation, regulations, cooperation, mobility, and storage.

One of the principal pillars of "Andean wisdom" (Regalski 1994) and of the resilience of Andean agriculture and rural livelihoods is the *Complementaridad*, a complementary system of multiple resource reliance, of agricultural strategies and economic and social activities (Murra 1975, Regalski & Calvo 1989). A second major ethical, economic and socio-cultural concept of Andean communities is that of *Reciprocidad*, a complex system of reciprocal obligations and forms of assistance of each person within a rural *Comunidad*:

> "[Reciprocity] must be seen as an element in the socio-cultural strategies Andean farmers use in attempting to reshape the market economy, based on the ethical principles of solidarity, equity, and a minimum degree of ecological sustainability. This is superior to mere maximization of individual benefits". (Rist 2000: 315)

In being guided by the principles of complementarity and reciprocity, agro-pastoral activities (Coppock & Valdivia 2001), the prudent use of the land and water resources and a pooling of knowledge, responsibilities, labor, at times also of land and other assets

Table 2   Resilience of Andean farming communities

FARMING COMMUNITIES	AGRICULTURAL LAND USE
- SABER ANDINO	- RISK MINIMIZATION
- RECIPROCIDAD	- SUSTAINABLE FORMS OF AGRICULTURAL LAND USE
- CAPACITACIÓN	
- ENABLEMENT, ENTITLEMENT EMPOWERMENT	- STRENGTHENING OF SUCCESSFUL TRADITIONAL CROPS, FORMS OF LAND CULTIVATION AND LAND USE
- ATTITUDES OF JUSTICE & EQUITY	
- AUTOGESTIÓN	- DEVELOPMENT OF NEW 'NICHES' OF SUSTAINABLE AND PROFITABLE AGRICULTURAL LAND USE, AND OF 'NICHE' PRODUCTS FOR MARKETS
- FOSTERING OF LOCAL LEADERSHIP	
- COMMUNAL COHERENCE	
- ENVIRONMENTAL PROTECTION	- FOSTERING OF COMPATIBLE FORMS OF AGROFORESTRY
- STRENGTHENING OF ECONOMIC BASE	
- SUSTAINABLE RESOURCE USE	- IMPROVEMENT OF IRRIGATION POTENTIAL AND PEST CONTROL
- IMPROVED RURAL INFRASTRUCTURES AND SERVICES	- PROMOTION OF SUCCESSFUL LAND USE INNOVATIONS & TECHNOLOGIES
- IMPROVED MARKET ACCESS	
- ENHANCEMENT OF HUMAN SECURITY AND QUALITY OF LIFE	

have supported the survival of a multi-faceted and resilient small-scale agriculture and of the viability of peasant communities in the Andes. Table 2 lists in a summarizing fashion the different components of traditional and potentially new forms and strategies of agricultural land use and of farming communities to strengthen and promote the resilience of Andean agriculture and rural livelihoods. Most experts agree today that this concept of Andean resilience should form the basis and core of rural and regional development initiatives. Andean resilience is not only based on a preservation of biodiversity and promotion of the cultural heritage and traditional approaches and strategies (Pohle 2004), but always has to cope with and adapt to changes and new developments. These changes may occur within a specific region, my be initiated by local actors and local and regional processes, and may require local responses and adaptations. Andean resilience is not only based on a preservation and promotion of the cultural heritage and traditional approaches and strategies, but it has also to cope with and adapt to changes and new developments. These changes may occur within a specific region, may be initiated by local actors and local responses and adaptations.

Today, the rural Andean regions are more closely integrated into national economies and societies, and the livelihoods, landscapes, and forms of agricultural systems, are increasingly affected by global influences, a process for which Bebbington (2001) has introduced the term of "Globalized Andes". These changing socio-economic conditions and forces bring to Andean farmers new challenges and at times additional problems. Hopefully though, they may offer to them new opportunities and agricultural possibilities that will enhance their rural livelihoods.

Unquestionably, Andean agriculture although rooted in heritage and tradition, has never been stable and immutable but has been dynamic in nature (Stadel 2003a, 2003b). This has been a necessity in response to environmental changes, land degradation and other hazards or even disasters. But it has also been a consequence of regional population increases or decreases and in- and outmigration, changing land tenure and land prices, of agricultural and technical innovations in seed and breeding standards, of pest and disease control, measures of irrigation and changing cropping and pastoral systems, of accessibility and transportation facilities to markets, as well as of external political and economic impacts. It appears though that through an effective and ubiquitous penetration of the Andean rural realm by outside national and international forces, processes and stakeholders, the pace of agricultural and rural changes has accelerated. Table 3 summarizes the most important recent changes of socio-economic conditions which have an impact on Andean agriculture and livelihoods. Bebbington (2000) refers to these changes as "livelihood transitions" and "place transformations". This has a tendency to reinforce the polarization of agricultural and rural livelihoods. On the one side, we find the new "cores" of a large-scale, capital-intensive and export-oriented agriculture, especially of flower-, fruit- and vegetable farms, often owned and operated by outside persons. The major objective of these large operations is a profit maximization which is pursued by high investments in implements, handling and transportation facilities, research, technology and marketing. This is often achieved at the detriment of environmental sustainability and a careful use of the water and land resource potential, at times also by an unacceptable exploitation of local workers. While the environment and the traditional local food basis may be "eroded" by these agricultural operations,

Table 3 Changing socio-economic conditions

INCREASED MARKET AND PROFIT ORIENTATION
- CONVERSION OF LAND USE FROM SUBSISTENCE ORIENTATION TO MARKET ORIENTATION WITH THE CULTIVATION OF EXPORT-ORIENTED PRODUCTS
- INTENSIFICATION OF AGRICULTURAL LAND USE, OFTEN WITH INCREASED APPLICATION OF IRRIGATION, IMPLEMENTS AND MECHANIZATION
- OFTEN AGRICULTURAL LAND AMALGAMATIONS BY PRIVATE LAND OWNERS, COOPERATIVES, CORPORATIONS AND GOVERNMENTS
- POTENTIALLY DETERIORATING LAND USE AT THE OUTSKIRTS OF THE COMMERCIAL FARMS

STRENGTHENING OF COMMUNITY SUBSISTENCE AGRICULTURE WITH REGIONALMARKET ORIENTATION
- SUSTAINABLE, TO THE LOCAL CONDITIONS ADAPTED LAND USE
- AGRICULTURAL SYSTEMS AND TECHNIQUES BASED ON PROVEN TRADITIONAL PRACTICES AND ON ECOLOGICALLY AND CULTURALLY ADAPTED METHODS
- AGRICULTURAL LAND USE PROTECTING ENVIRONMENTAL INTEGRITY AND CAUTIOUS RESOURCE USE (LAND, SOIL, WATER, VEGETATION)
- EMPHASIS ON NATIVE CROPS AND CONSERVATION OF BIODIVERSITY
- PRODUCTION AND PROCESSING OF NICHE PRODUCTS FOR UP-SCALE MARKETS

CHALLENGING NEW INTERFACES BETWEEN CONSERVATION AND AGRICULTURE/FORESTRY
- CONFLICTING OR COMPLEMENTARY INTERESTS AND RESULTING LAND USE
- TRANSITION/BUFFER ZONE BETWEEN AGRICULTURAL LAND USE AND BIOSPHERES, BLENDING AGRICULTURAL PRODUCTION/FOREST USE AND CONSERVATION, OR SHARPLY DELINEATED EXCLUSIVE ECONOMIC AND CONSERVATION ZONES?

URBAN GROWTH AND RURAL INVASION
- INVADING URBAN LAND USE WITH TENTACLES OF LAND SPECULATION. INTENSIFICATION OF URBAN MARKET ORIENTED HORTICULTURE, OR 'BLIGHTED' AGRICULTURAL LAND
- CONVERSION OF AGRICULTURAL LAND TO TRANSPORTATION INFRASTRUCTURES (HIGHWAYS; AIRPORTS), SHOPPING CENTERS; INDUSTRIAL PARKS, RECREATION COMPLEXES
- LAND USE CHANGES RESULTING FROM FOREIGN REMITTANCES

ALTERNATIVE RURAL ECONOMIC ACTIVITIES
- ECOTOURISM WITH A CONSERVATION OF ECOLOGICALLY IMPORTANT OR VULNERABLE AREAS COMBINED WITH ENVIRONMENTALLY COMPATIBLE TOURISTIC INFRASTRUCTURE
- AGRO TOURISM WITH A MAINTENANCE OF AGRICULTURAL LAND USE COMBINED WITH "SOFT" TOURISTIC INFRASTRUCTURES

Compiled by C. Stadel

this development is often sanctioned or promoted by national governments for the sake of "agricultural progress" and national development goals. On the other side of the spectrum of agricultural operations and rural livelihoods remain the prevailing *Minifundios* and family farmsteads, predominantly oriented towards subsistence farming or a production for regional markets. These "unseen" and "unknown" (CHAMBERS 1983: 23) communities remain generally weak, powerless, isolated, and have been largely screened from outside interference or assistance. Today though, there appears to be a greater awareness of and concern for the vulnerable or marginalized rural environments and agricultural communities. Research projects, international and national development programs and projects of many non-governmental organizations focus on "pro-poor" objectives and strategies and addressing "local geographies of poverty" (ANNIS & HAKIM 1988, KLAGGE 2002, LAURIE 2007, MILBOURNE 2004).

Pleading for a *Campesino*-Oriented Development

In view of the ever increasing influence of external agents and processes and the widespread impact of globalization in various forms, it appears paramount to counterbalance these influences and forces by local voices and visions (RHOADES 2000). Based on his experience of a multi-year research and development partnership with indigenous communities in the Cotacachi region of northern Ecuador, RHOADES (2006) pleads for a "Development with Identity", e. g. for a genuine participatory development process based on the wisdom of ancestral knowledge, on traditional livelihoods, agricultural systems, on self-determination, and proven forms of social organization. Ideally, this locally – based research and development agenda can be complemented by a genuine partnership with an external scientific, technical and financial input. This could help to enhance the awareness of the local population for the complex local-global issues and better prepare it for adaptive measures which hopefully will create new opportunities for farmers and rural communities (NEUBERT & MACAMO 2002).

The following conceptual model (Fig. 4) is an attempt to portray the different aspects of a *campesino*-oriented development approach. This model and the following development guidelines take into account – albeit in a tentative and not complete form – the perceptions, objectives and strategies of Andean researchers and practitioners (STADEL 2000: 61–62). In summary, a *campesino*-based development should be based on:

- an understanding and a valorization of the *saber campesino*.
- a harmony between the environment and society (*cosmovisión andina*).
- the potential and limitations of the local physical and human environments.
- the principles of long-term economic viability for the local communities, on a respect for cultural values and traditions, on social well-being, equity and justice for all segments of the society, and on the maintenance or restoration of environmental quality.
- the traditions of community organization and the principles of economic and social *reciprocidad* (communal solidarity and cooperation).
- the priorities for an environmentally compatible agriculture (*agroecología*), biodiversity, and for cultural diversity.

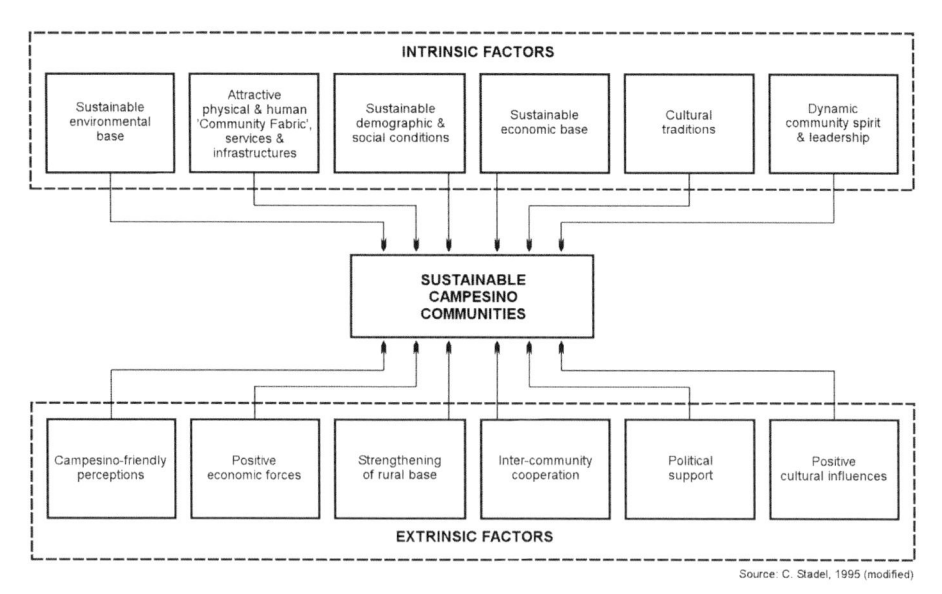

Fig. 4 Sustainable campesino communities

- a complementarity between local traditions and strategies and compatible external innovations.
- the requirements for a long-term economic, social, and political security, stability, and harmony for families and communities.
- locally perceived needs, priorities, methods, strategies, and techniques.
- the principles that development efforts have a priority focus on the most fragile environments and on the most marginal segments of the population.
- the philosophy that *campesinos* are not subjects who have to adapt themselves to external concepts and demands, but are participants and genuine partners in the development process.

General Observations on Global Impacts on Andean Agriculture and Rural Livelihoods and Tentative Guidelines for Andean Research

At the "Symposium on Climate Change: Organizing the Science in the American Cordillera" in Mendoza, Argentina (CONCORD 2006: 31), the author addressed potential changes and adaptations of agricultural land use and of rural livelihoods under the impact of climate change and socio-economic globalization. The anticipated warming trend may result in upward shifts of some altitudinal zones of crops and pastures and of the upper limits of settlements, and it may shorten the frost hazards at intermediate levels. Changing precipitation regimes may also affect the rural water supply and irrigation potential, most notable perhaps resulting in the melting of the snow or ice caps of the highest summits. This has been recently documented in the Cotacachi region of Ecuador where the 4937 m high Nevado (!) Cotacachi has lost its cap of glaciers and permanent snow over the last few decades, a fact which has severe impacts on the water availability of the region (RHOADES 2006: 64–73). Oscillating climate conditions with significant repercussions on agriculture and rural livelihoods occur during *"El Niño"* or *"La Niña"* events with their altered regional precipitation regimes often requiring short-term adaptations of agricultural activities.

In the current focus of the scientific debate on climatic changes, the impacts of changing socio-economic conditions largely resulting from external – at times global-level – forces and actors should not be overlooked. Overall, agricultural activities and rural livelihoods have been greatly affected by an increased market and profit orientation. While the impacts of these changes may be ambivalent, these new developments have often widened the socio-economic gap within specific rural regions. Reacting to these potential threats to the viability and autonomy of regional economies, may communities have attempted to protect their regional environmental base and to safeguard a sustainable small-scale agriculture.

The new awareness of environmental protection, the delicate balance between agriculture and forestry on the one hand and conservation or environmental rehabilitation on the other hand, represents new challenges and opportunities. In some areas, the interface between ecological postulates and regional economic necessities has stimulated the development of various forms of ecotourism or agro-tourism with their emphasis on environmental protection and sustainable small-scale agriculture. Undoubtedly, one of the most fundamental factors of the socio-economic transformations of most rural Andean regions has been the improvement of the transportation infrastructure and the enhanced or new forms of communication. This has resulted in various forms of interactions between regional cities, metropolitan centers and international destinations and the rural Andean communities. For instance, family members may work periodically or permanently in the larger cities of the country or abroad, may send financial remittances to their villages, or may culturally bring city life of foreign tastes and habits to their homeland.

Suggestions for Research and Development Priorities in the Rural Andes

Based on the observations and experiences in various parts of the tropical Andes, the following general conclusions relating to Andean agriculture and rural livelihoods and suggestions for research and development priorities may be made – albeit in a tentative fashion. These comments and recommendations do not suggest that they may not have been properly considered in research, policies and strategies. They are meant to provide a summary of observations and experiences of Andean scholars and practitioners:

- Given the heterogeneity and diversity of Andean environments and community, any generalizations about the *campesinado* of the tropical Andes, about the nature and extent of its vulnerability and the various forms of resilience and adaptation appear to be problematic.
- Global changes are affecting all regional levels and all segments of Andean communities. The global changes may however have varied impacts on specific regions and people.
- While global climatic changes have major repercussions on human livelihoods, other types of global changes (cultural, social, economic and political ones) have

also a major impact on environments and livelihoods. Furthermore, changes affecting the rural Andes are not only of a "global" nature, some or them emanate at a regional or national level.

- Research on global changes should be carried out on the basis of a genuine cross-disciplinary integrative approach.
- Farming communities should be empowered to successfully confront the changes and exterior challenges on the basis of an enhanced self-confidence, pride, knowledge and autonomy. They should be encouraged to "screen" the external influences through local ecological and socio-cultural "filters".
- While there is a persistent need for research on local regional case studies, the research on global changes should also be focused on entire watersheds, ecological belts and niches, cultural realms, and in general on highland-lowland interaction systems.
- In view of a pervasive penetration of market orientation, corporate economic interests, modernization and globalization, research on Andean land use systems should focus on alternative and diversified forms of agriculture, e. g. agro-ecology, agro-forestry, agro-biodiversity, agro-tourism, agro-industry.
- Andean farming communities should be informed about good agricultural and livelihood practices. This "pooling" of successful experiences and their promotion could enhance further adoption, replicate successes and catalyze the development process in wider regions and larger *campesino* communities.
- Research on the rural Andes should focus both on fragile and vulnerable environments, as well as on successful "islands of sustainability".

Concluding Remarks

The experiences of scientists and practitioners confirm that the environmental and social compatibility of agricultural activities and rural livelihoods and of development programs form the basis for a sustainable future of the Andes (STADEL 2006). It has also become obvious that ecological considerations, parameters and policies cannot be isolated from the cultural and social context of the Andes and that there is a close interrelationship between the Andean environment and Andean livelihoods (LLAMBI et al. 2005).

This paper emphasized the diversity of agricultural landscapes and systems of the tropical Andes and attempted to portray the major challenges and constraints for farming and rural living. But it also focused on the wealth and potential of natural and human resources and on the long tradition of the *Saber Andino* which, over the centuries, gave witness to many successful forms of resilience and adaptation to environmental, economic and social challenges and changes (APFFEL-MARGLIN 1998, LAURIE et al. 2005, RIST 1993). While many opportunities and problems are local in nature and may require local responses, increasingly today rural environments and communities are affected by national and global processes and actors, and may therefore require strategies at a wider regional, national, or even global level (GSAENGER 1998). Yet, it appears paramount, that the stewardship over the Andean environment and resources should be kept in the responsibility and control of indigenous or local communities (Hoffmann 2007). Furthermore, any conservation or development initiatives must be based on the cultural heritage, the needs be based on the cultural heritage, the needs and priorities of local populations. In the light of neo-liberal market orientation and growing global impacts, the voices of the rural populations should be heard and their own livelihoods secured in a sustainable fashion Flora et al. 1997, Gibbs 2005, Gregory 2000). It is therefore with considerable hesitation that this overview of Andean rural environments, their potentials and problems has been formulated in a rather generalizing fashion. Also, the proposed guidelines for social research in the context of global changes within the Andean realm have to be quite tentative but could hopefully make a contribution to Andean research and development approaches.

In conclusion, the resilience of Andean rural communities has to rely on the dual pillars of a wise adherence to proven traditional values and practices and on the local "voices" and actions, as well as on locally controlled and sustainable forms of adaptations in a genuine cooperation with external actors. This careful balance – in the past and in the present – has at times not successfully been achieved and has either led to forms of stagnation and marginalization, or to an uncontrolled invasion of poorly adapted economic, technological and cultural influences which may have resulted in deep-rooted crises and disparities. Yet, the growing local *conscientización*, *autogestión*, "stewardship" for the environment and the resources and empowerment of local stakeholders should in many cases attenuate the ecological and social vulnerabilities and strengthen the potential of resilience and of successful and sustainable adaptations to new developments.

References

Allan, N. J. R. (1986): Accessibility and altitudinal zonation models of mountain. Mountain Research and Development. Vol. 6 (3): 185–194.

Annis, S.; P. Hakim (eds.) (1988): Direct to the Poor. Grassroots Development in Latin America. Boulder and London.

Apffel-Marglin, F. (ed.) (1998): The spirit of regeneration: Andean culture confronting Western notions of development. London.

Bebbington, A. (1997): Social capital and rural intensification: local organizations and islands of sustainability in the rural Andes. The Geographical Journal 163(2): 189–197.

Bebbington, A. (2000): Reencountering Development: Livelihood transitions and place transformations in the Andes. Annals of the Association of American Geographers 90: 495–520.

Bebbington, A. (2001): Globalized Andes? Livelihoods, Landscapes and Development. Ecumene 8(4): 414–436.

Bohle, H:-G. (2008): Leben mit Risiko – *Resilience* als ein neues Paradigma für die Risikowelten von morgen. In: Felgentreff, C.; T. Glade (eds.) Naturrisiken und Sozialkatastrophen. Berlin/Heidelberg: 435–441.

Brush, S. B. (1987): Diversity and Change in Andean Agriculture. In: Little, P. D. et al. (ed.): Lands at Risk in the Third World. Local-level Perspectives. Boulder/London: 271–289.

Chambers, R. (1983): Rural Development. Putting the Last First. Harlow.

CONCORD (ed.) (2006): Symposium on Climate Change: Organizing the Science in the American Cordillera. Abstracts. Mendoza.

COPPOCK, D. L.; C. VALDIVIA (ed.) (2001): Sustaining Agropastoralism on the Bolivian Altiplano: The case of San José Llanga. Logan.

DELGADO, B. F. (1993): La agroecología en las estratégias del desarrollo rural (una experiencia institucional). Trabajo del Colegio Andino 9 Cuzco.

FLORA, C. et al. (1997): Negotiating participatory action research in an Andean Ecuadorian sustainable agriculture and natural resource management program: Practicing Anthropology 19: 20–25.

GADE, D. (1999): Nature and Culture in the Andes. Madison.

GIBBS, D. (2005): Exploring local capacities for sustainable development. Geoforum 36: 407–409.

GREGORY, R. (2000): Using stakeholder values to make smarter environmental decisions. Environment 42: 34–44.

GSAENGER, H. (1998): Capacity-Building for Rural Development at the Micro-, Meso- and Macro-Levels. Journal für Entwicklungspolitik 11(2): 137–177.

HEWITT, K. (1988): The Study of Mountain Lands and People: A Critical Overview. In: ALLAN, N. J. R.; G. KNAPP; C. STADEL (eds.): Human Impact on Mountains. Totowa, N.J.: 6–23.

HOFFMANN, D. (2007): The Sajama National Park in Bolivia. A Model for Cooperation among State and Local Authorities and the Indigenous Population. Mountain Research and Development 27(1): 11–14.

HOLLING, C. S. (1973): Resilience and stability of ecological systems, Annual Review in Ecology and Systematics 4: 1–23.

IVES, J. D. (1986): Editorial, Mountain Research and Development 6(3): 183–184.

JANSSEN, M. A., E. ORSTROM (2006): Resilience, Vulnerability and Adaptation, IHDP Newsletter 1: 10–11.

KLAGGE, B. (2002): Lokale Arbeit und Bewältigung von Armut – eine akteursorientierte Perspektive. Geographische Zeitschrift 90(3+4): 194–211.

KNAPP, G. (1991): Andean Ecology. Adaptive Dynamics in Ecuador. Boulder/San Francisco/Oxford: Dellplain Latin American Studies 27.

LAURIE, N. (2007): Introduction: How to dialogue for pro-poor water. Geoforum 38: 753–755.

LAURIE, N. et al. (2005): Ethnodevelopment: Social Movements. Creating Experts and Professionalizing Indigenous Knowledge in Ecuador Antipode 37(3): 470–496.

LLAMBI, L. D. et al. (2005): Participatory Planning for Biodiversity Conservation in the High Tropical Andes: Are Farmers Interested? Mountain Research and Development 25(3): 200–2005.

MILBOURNE, P. (2004): The local geographies of poverty: a rural case study, Geoforum 35: 559–575.

MURRA, J. V. (1975): El control vertical de un máximo de pisos ecológicos en la economía de las sociedades andinas. In: *Instituto de Estudios Peruanos* (eds.): Formaciones económicas y politicas del mundo andino. Lima, 59–115.

NEUBERT, D., E. MACAMO (2002): Entwicklungsstrategien zwischen lokalem Wissen und globaler Wissenschaft. Geographische Rundschau 54(10): 12–17.

REGALSKI, P. A. (1994): La sagesse des Andes. Une expérience originale dans les communautés andines de Bolivie. Geneva.

REGALSKI, P. A.; L. M. CALVO (1989): La Plasticidad del Manejo Andino y sus desafios. Cochabamba.

Resilience Alliance. www.resalliance.org

RHOADES, R. E. (2000): Integrating Local Voices and Visions into the Global Mountain Agenda. Mountain Research and Development 20(1): 4–9.

RHOADES, R. E. (ed.) (2006): Development with Identity. Community, Culture and Sustainability in the Andes. Oxfordshire: CABI Publishing.

RHOADES, R. E.; S. I. THOMPSON (1975): Adaptive strategies in alpine environments: Beyond ecological particularism. American Ethnologist 2: 535–551.

RIST, S. (2000): Linking Ethics and the Market: Campesino Economic Strategies in the Bolivian Andes. Mountain Research and Development 20(4): 310–315.

RIST, S. (2007): The importance of Bio-Cultural Diversity for Endogeneous Development. In: HAVERKORST, B.; S. RIST (eds.): Endogeneous Development and Bio-Cultural Diversity. The interplay of worldview, globalization and locality. COMPAS series on Worldviews and Sciences 6. Berne-Leusden: 76–81.

STADEL, C. (1991): Environmental stress and sustainable development in the Tropical Andes. Mountain Research and Development 11(3): 213-223.

STADEL, C. (1992): Altitudinal belts in the tropical Andes: their ecology and human utilization'. In: MARTINSON, T. (ed.), Benchmark 1990. Conference of Latin Americanist Geographers 17/18: 45-60.

STADEL, C. (1995): Perzeption des Umweltstresses durch Campesinos in der Sierra von Ecuador. In: MERTINS, G.; W. ENDLICHER (eds.): Umwelt und Gesellschaft in Lateinamerika (Marburger Geographische Schriften 129): 244–265.

STADEL, C. (2000): Development and Sustainability in Latin America. In: BORSDORF, A. (ed.): Perspectives of Geographical Research on Latin America for the 21st century. ISR-Forschungsbericht, Vienna, 54–70.

STADEL, C. (2001): „Lo Andino": andine Umwelt, Philosophie und Weisheit. Innsbrucker Geographische Studien 32: 143–154.

STADEL, C. (2003a): Indigene Gemeinschaften im Andenraum – Tradition und Neu-orientierung. In: EITEL, B. et al. (ed.): Naturrisiken und Naturkatastrophen – Lateinamerika (HGG – Journal 18). Heidelberg: 75–88.

STADEL, C. 2003b: L'agriculture andine: traditions et mutations. In: CERAMAC: Crises et mutations des agricultures de montagne. Clermont-Ferrand: 193–207.

STADEL, C. (2006a): Entwicklungsperspektiven im ländlichen Andenraum. Geographische Rundschau 58(10): 64–72.

STADEL, C. (2006b): Report on the Latin American Working Group. In: *UNESCO* (ed.): GLOCHAMORE. Projecting Global Change Impacts and Sustainable Land Use and Natural Resource Management in Mountain Biosphere Reserves. Paris: 167-174.

TAPIA, M. E. (2000): Mountain Agrobiodiversity in Peru. Seed Fairs, Seed Banks, and Mountain-to-Mountain Exchange. Mountain Research and Development 20(3): 220–225.

THOMAS, R. B.; B. EINTERHADER; S. D. MCRAE. (1979): An anthropological approach to human ecology and adaptive danamics. Yearbook of Physical Anthropology 22: 1–46.

VOGL, C. R. (1990): Traditionelle andine Agrartechnologie. Beschreibung und Bedeutung traditioneller landwirtschaftlicher Techniken des peruanischen Andenraums aus der Sicht ökologischer und sozio-ökonomischer Anpassung. Wien (Thesis).

ZIMMERER, K. S. (1999): Overlapping Patchworks of Mountain Agriculture in Peru and Bolivia: Toward a Regional-Global Landscape Model. Human Ecology 27 (1): 135–165.

Indigenous and Local Concepts of Land Use and Biodiversity Management in the Andes of Southern Ecuador

Perdita Pohle[1]

Abstract

In southern Ecuador, a region of heterogenic ethnic, socio-cultural and socio-economic structures, profound knowledge of ethnic-specific human ecological parameters is crucial for the sustainable utilization and conservation of tropical mountain forests. In order to satisfy the objectives of environmental protection on the one hand and the utilization claims of the local population on the other hand, detailed analysis of human ecological parameters is needed. This article aims to provide an introduction to the indigenous and local population of the area surrounding Podocarpus National Park in southern Ecuador. In the case of the indigenous Shuar (lowland Indians) and Saraguros (highland Indians) and the local mestizos, fundamental differences occur not only in the attitudes towards the tropical rainforest and the management of forest resources but also in wider economic and social activities, including all strategies for maintaining their livelihood.

1 Introduction

By now it is sufficiently well understood that any attempt to preserve primary forest in the tropics is destined to fail if the interests and use claims of the local population are not at the same time, and in the long term, taken into account. Therefore, in addition to strict protection of the forests, an integrated concept of nature conservation and sustainable land use development needs to be sought (e. g. ELLENBERG 1993). The DFG-research project presented here will figure out the extent to which traditional ecological knowledge and indigenous biodiversity management strategies can be made available for a long-term land use development. The project chose to use a specialized approach, namely to investigate 'Indigenous/ Local Knowledge Systems' as part of the ethno-ecological methodology (e. g. MÜNZEL 1987, POSEY & BALÉE 1989, WARREN et al. 1995, MÜLLER-BÖKER 1999, NAZAREA 1999). In biodiversity-rich places local people usually have a detailed ecological knowledge, for example, of species, ecosystems, ecological relationships and historical or recent changes of them. Numerous case studies have shown, how traditional ecological knowledge and traditional practices serve to effectively manage and conserve natural and man-made ecosystems and the biodiversity contained within (e. g. POSEY 1985, TOLEDO et al. 1994, BERKES 1999, FUJISAKA et al. 2000, POHLE 2004). In ongoing interdisciplinary and integrative research projects like BIOTA AFRICA (Biodiversity Monitoring Transect Analysis in Africa, German Federal Ministry of Education and Research), STORMA Indonesia (Stability of Rainforest Margins in Indonesia, German Research Foundation, Collaborative Research Centre 552) or within the interdisciplinary programme of the National Centre of Competence in Research North-South implemented by the Swiss National Science Foundation

---

[1] Institut für Geographie, Universität Erlangen-Nürnberg, D-91054 Erlangen; ppohle@geographie.uni-erlangen.de.

(SNSF), investigations on traditional ecological knowledge and biodiversity management are an integral part.

2 The Tropical Mountain Rainforests of Southern Ecuador – a "Hot Spot" of Biodiversity

The tropical mountain rainforests of the eastern Andean slopes in southern Ecuador have an extraordinary rich biodiversity (BARTHLOTT et al. 1996). The area under study, the Podocarpus National Park and its surroundings, is especially noteworthy for its species diversity and belongs to one of the so-called "hot spots" of biodiversity worldwide (MYERS et al. 2000). The tropical mountain rainforests of southern Ecuador are of crucial importance for the preservation of genetic resources, and play an important role as an ecosystem and habitat for flora and fauna. At the same time humans have lived here and sustain themselves since centuries. However, in more recent times (the past four or five decades), these mountain forest ecosystems, which have been described as particularly sensitive (Die Erde 2001), have come under enormous pressure from the expansion of agricultural land – especially pastures –, the extraction of timber, the mining of minerals, the tapping of water resources and other forms of human intervention. According to Hamilton et al. (1995), 90 % of the original forest cover in the Andes can be regarded as either destroyed or altered. At present, in Ecuador the annual deforestation rate of 1.7 % is the highest of all South American countries (*FAO* 2006).

3 Indigenous and Local Ethnic Communities

The northern and eastern surroundings of the Podocarpus National Park, the special areas under study, are settled by indigenous Shuar and Saraguro communities as well as Mestizo- Colonos (cf. POHLE 2008, fig. 1).

The **Shuar** area of settlement extends from the lower levels of the tropical mountain rainforest (approx. 1,400 m a. s. l.) down to the Amazonian lowland *(Oriente)* in the region bordering Peru. The Shuar, who belong to the Jívaro linguistic group (Amazonian Indians), are typical forest dwellers who practice shifting cultivation mainly in subsistence economy. Their staple crop is manioc which they plant together with taro and plantains on small rotating plots in forest gardens. They also fish, hunt and gather forest products. During the past decades some Shuar have also begun to raise cattle and some are engaged in timber extraction as well. With the infrastructural opening of several traditional Shuar territories, the way of life and the livelihood of many Shuar have dramatically changed during the past decades (cf. RUDEL & HOROWITZ 1993).

The **Saraguros**, highland Indians who speak Quichua, live as agro-pasturalists for the most part in the temperate mid-altitudes (1,800-2,800 m a. s. l.) of the Andes *(Sierra)* in southern Ecuador. As early as the 19th century the Saraguros kept cattle to supplement their traditional "system of mixed cultivation", featuring maize, beans, potatoes and other tubers (GRÄF 1990). Now, cattle ranching has developed into the main branch of their economy. Like other indigenous groups of Ecuador, the Saraguros have undergone cultural change: most Saraguros are today Catholic Christians who hardly speak any Quichua outside their main area of settlement, but only Spanish, and increasingly seek out job opportunities away from agriculture (BELOTE 1998).

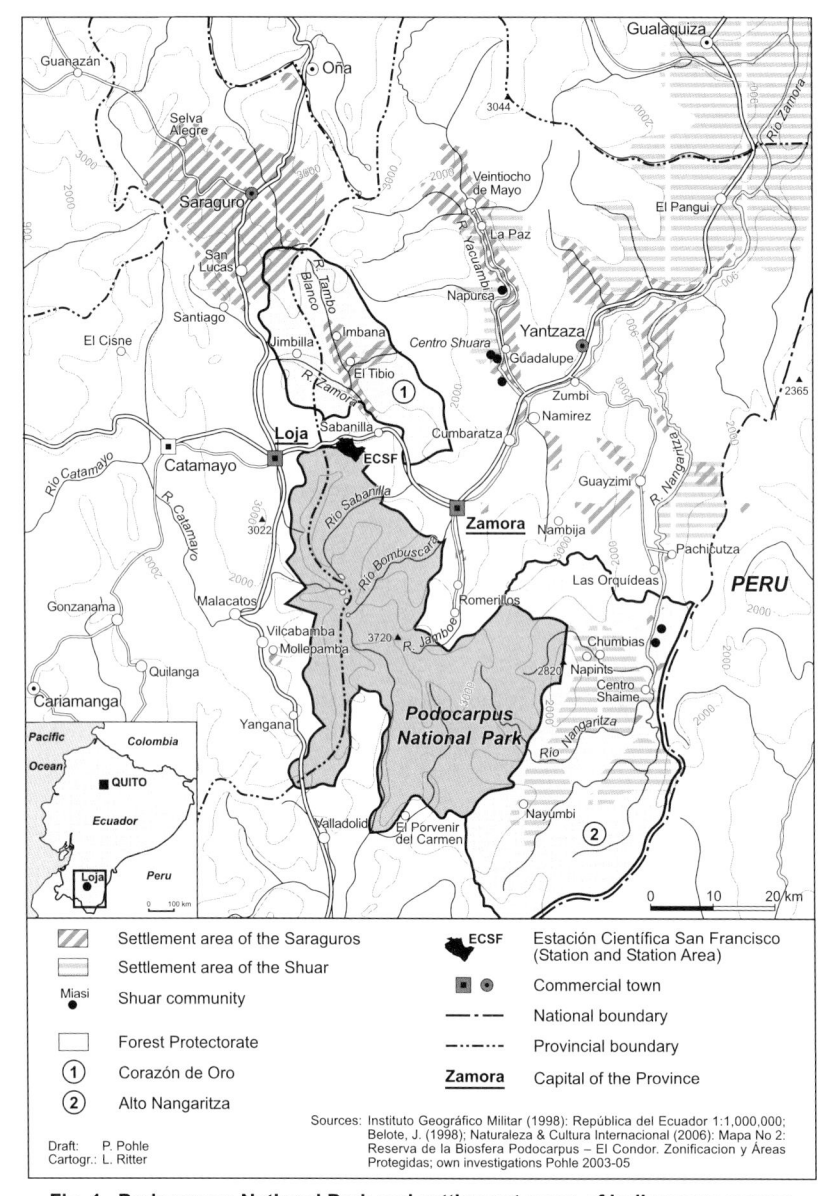

Fig. 1 Podocarpus National Park and settlement areas of indigenous groups

Even though the traditional forms of life and livelihood, both of the Shuars and the Saraguros, have greatly changed under the pressure of external influences (e. g. missionary activities, agrarian colonization), both indigenous groups have been able to preserve core areas of their traditional culture, including an extensive specialized knowledge of their environment, along with numerous life-support strategies through natural resource management.

In the North and East of Podocarpus National Park *Mestizo-Colonos* are the most dominant ethnic group in numbers. They are colonists of mestizo ethnicity, who came into the area during the past four or five decades to log timber and to practise cattle farming and agriculture. Their migration was especially encouraged by the national land reform of 1964, which opened to poor peasants the possibility of becoming landowners by clearing and transforming forest into pastures and fields. A severe drought in 1968 in Loja province (VAN DEN EYNDEN 2004) boosted this colonization process. Additionally, the colonization was facilitated by the construction of the Loja-Zamora road (1962) connecting the Andean and Amazonean areas.

4   Aims and Methods of the Ethnoecological Research Project

The ethnoecological research project was carried out within the DFG-Research Unit FOR 402: "Functionality in a Tropical Mountain Rainforest: Diversity, Dynamic Processes and Utilization Potentials under Ecosystem Perspectives" (www.bergregenwald.de). Within the ethnoecological project ethnicity is viewed as a driving factor in the relationship between man and his environment. Fundamental differences between the indigenous Shuar and Saraguro communities as well as the local Mestizo-Colonos occur not only in attitudes towards the tropical rainforest and the management of forest resources but also in wider economic and social activities, including all strategies for maintaining livelihood.

During 2004, 2005 and 2006 ethnoecological, especially ethnobotanical and agrogeographical research was undertaken in sample communities of the Shuar (Shaime, Napints, Chumbias), the Saraguros (El Tibio) and the Mestizo-Colonos (Sabanilla). The goals were:

- To document the indigenous and local knowledge of traditionally utilized wild and cultivated plants (the ethnobotanical inventory was undertaken following the "Code of Ethics");
- To analyze current forms of land use including the cultivation of forest and home gardens;
- To evaluate ethno-specific life-support strategies as well as strategies for natural resource management.

5   The Significance of Plant Use for the Shuar, Saraguros and Mestizo-Colonos

In fig. 2 the number of plant species (wild and cultivated) used by the Shuar, the Saraguro and Mestizo-Colonos communities are listed. Plant uses were recorded according to categories of utilization like food, medicine, construction etc.

In biodiversity-rich places local people usually have a detailed ecological knowledge e. g. of species, ecosystems, ecological relationships and historical or recent changes to them (e. g. WARREN et al. 1995). This applies wholly to the Shuar communities. As traditional forest dwellers the Shuar of the Nangaritza valley have a comprehensive knowledge of plants and their utilization. All households make extensive use of forest

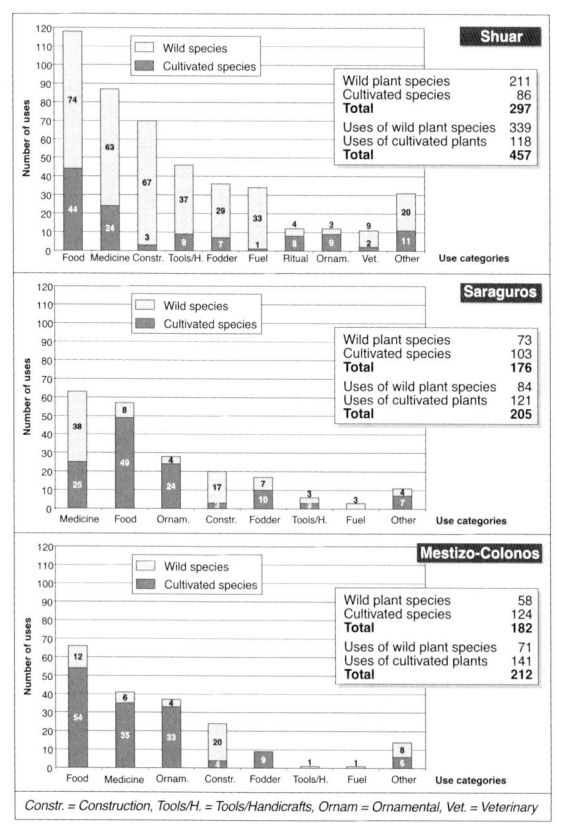

Fig. 2   Wild and cultivated plant species used by the Shuar (Chumbias, Napints, Shaime), the Saraguros (El Tibio) and the Mestizo-Colonos (Sabanilla) according to use categories. (Note: one species can be found in more than one use category.)

products. According to the ethnobotanical survey[2], the actual inventory of traditionally used wild plants of the Shuar includes 211 species. Most of the plants are used to supplement the diet (74). Given the lack of state health care, medicinal plants also assume great significance (63). Many plants, too, are used as construction material (67), as tools and for handicrafts (37), as fuel, fodder or as ritual plants. The Shuar use forest products exclusively for their own needs, there is virtually no commercialization.

The Saraguros from El Tibio have a far less comprehensive knowledge on wild plant species and their utilization. The actual ethnobotanical inventory includes only 73 wild plant species. Most of them are ruderal plants used as medicine (38) or plants used for their wood (17). As agro-pasturalists they have conversed most of the primary forest into pastures, home gardens and fields, leaving forest remains only along mountain ridges or in river ravines. Their actual plant knowledge reflects this traditional way of life. They have a comprehensive knowledge of cultivated plants (103) mainly pasture and crop plant species – even more than the Shuar (86) – but they are less familiar with

[2]   The ethnobotanical survey was conducted by Andrés Gerique (cf. GERIQUE & VEINTIMILLA 2007).

forest plant species. The latter knowledge is mainly reduced to woody varieties which they extract and sell before clearing the forest.

The Mestizo-Colonos of the surrounding areas of Sabanilla base their economy on cattle ranching. They have converted large areas of forests into pastures. As settlers from the western and most arid area of Loja Province they seem not to be familiarized with the local flora and hence make only little use of it. The actual ethnobotanical inventory – although not completed yet – includes a total of only 58 wild plant species (fig. 2). Timber is the main forest product (20), while some ruderal plants and the fruit of a few tolerated tree species are used as food (12) or as medicine (6). However, the Mestizo-Colonos cultivate 54 different species for food and 35 species for medicinal purposes and have a comprehensive knowledge of pasture species.

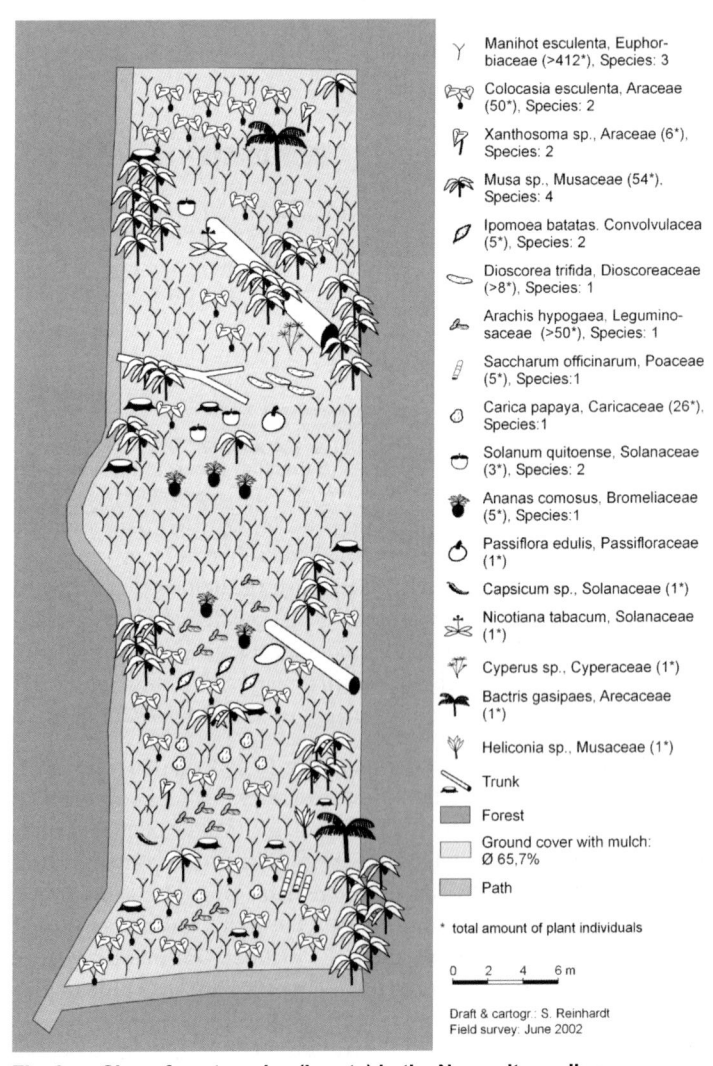

Fig. 3   Shuar forest garden (huerta) in the Nangaritza valley

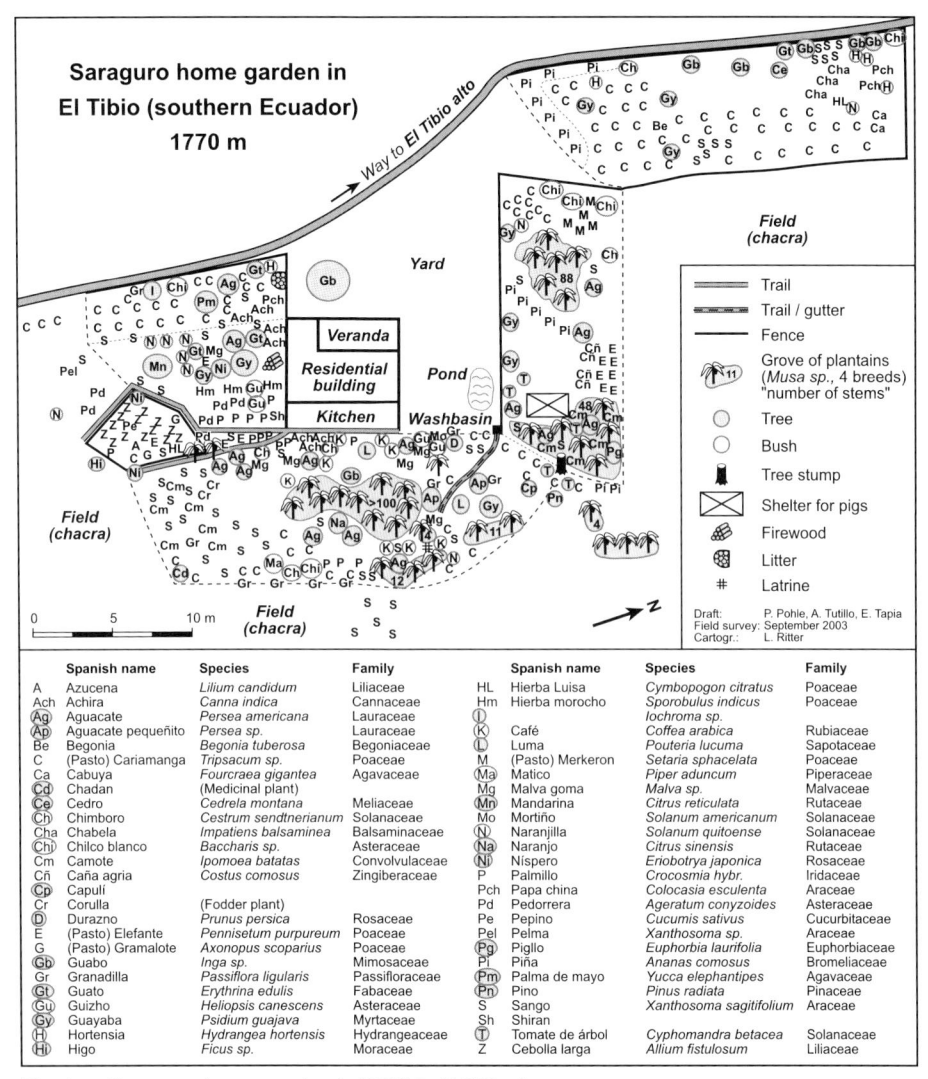

Fig. 4 Saraguro home garden in El Tibio (1,770 m)

6 Agrobiodiversity in Shuar and Saraguro Tropical Home Gardens *(huertas)*

The tropical home gardens of indigenous and local communities are generally regarded as places of great agrobiodiversity and refuges of genetic resources (WATSON & EYZAGUIRRE 2002). Furthermore, they contribute significantly to securing and diversifying food supplies. This applies wholly to Shuar and Saraguro gardens, which feature a large number of both wild and cultivated species and which play their part in providing subsistence needs. Staple crops, such as maize, tubers and beans, may be cultivated primarily in *chacras* (fields), but home gardens *(huertas)* have an essential role to play in supplementing the diet with fruits and vegetables, furnishing households with medicinal plants and spices, and fodder and timber.

The forest gardens of the Shuar (fig. 3) are characterized by an especially great diversity of species and breeds. In five *huertas* studied (size: approx. 600-1,000 m^2), a total of 185 wild and cultivated plant species and breeds were registered. For the most part they serve as nutritional items (58 %) or medicines (22 %). The main products cultivated are starchy tubers like manioc *(Manihot esculenta)* and taro *(Colocasia esculenta)*, along with various breeds of plantains *(Musa sp.)*. Moreover, the planting of a large number of traditional local breeds was documented: for example 29 breeds of manioc and 21 breeds of *Musa* sp. – a further indication of the crucial significance that home gardens have for the *in situ* conservation of botanical genetic resources (MÜNZEL 1989: 434).

The *huertas* of the Saraguros likewise display a great diversity of useful plants. In one sample home garden studied in El Tibio (fig. 4), 51 species of cultivated plants were identified. In total 95 cultivated plants were registered among the Saraguros of El Tibio. Again, the majority are plants that supply nutritional value (41 %), followed by medicinal and ornamental plants (each 20 %). The most important cultivated products are plantains, tubers and various types of fruit. Given their relatively dense and tall stands of trees, the multi-tiered arrangement of plants and the great diversity of species, the gardens of the Saraguros can be seen as an optimal form of exploitation in the region of tropical mountain rainforests.

Photo   A. Gerique

Photo 1   Napints (1000 m): scattered settlement of the Shuar in the tropical rainforest at the eastern periphery of the Podocarpus National Park

7   Indigenous Concepts of Biodiversity Management – their Contribution to a Sustainable Land Use Development

If the hypothesis is accepted that a multi-facetted economic and cultural interest in the forest on the part of indigenous and local communities offers effective protection against destruction, then a key role must be assigned to the analysis and evaluation of the ethno-specific knowledge about tropical mountain rainforests and their potential uses. Both indigenous groups have developed natural resource management strategies that could be used and expanded, in line with the concept "preservation through use", for future biodiversity management, but this should be done only in an ethno-specific way.

The Shuar traditional way of managing biodiversity is based on a sense of being closely bound culturally, spiritually and economically to the forest. The sustainability of their form of land use has long since been put to the test (MÜNZEL 1977, 1987). As traditional forest-dwellers, sustainable elements of biodiversity management can be found in (photo 1):

- Their regulated practice of shifting cultivation, which – given the correspondingly long time for regeneration – is thought to conserve the soil and the vegetation. The system of cultivation and fallow on small rotating plots (fig. 5) has much in common with ecological succession in that it uses the successional process to restore the soil and the vegetation after use for farming (KRICHER 1997: 179). In the Shuar forest gardens the fallow periods last for about 24-30 years while the cultivation periods covers 4 years.

- Their tending of forest gardens according to principles of agroforestry and mixed cropping with a high agrobiodiversity and a particular high breed variety of cul-

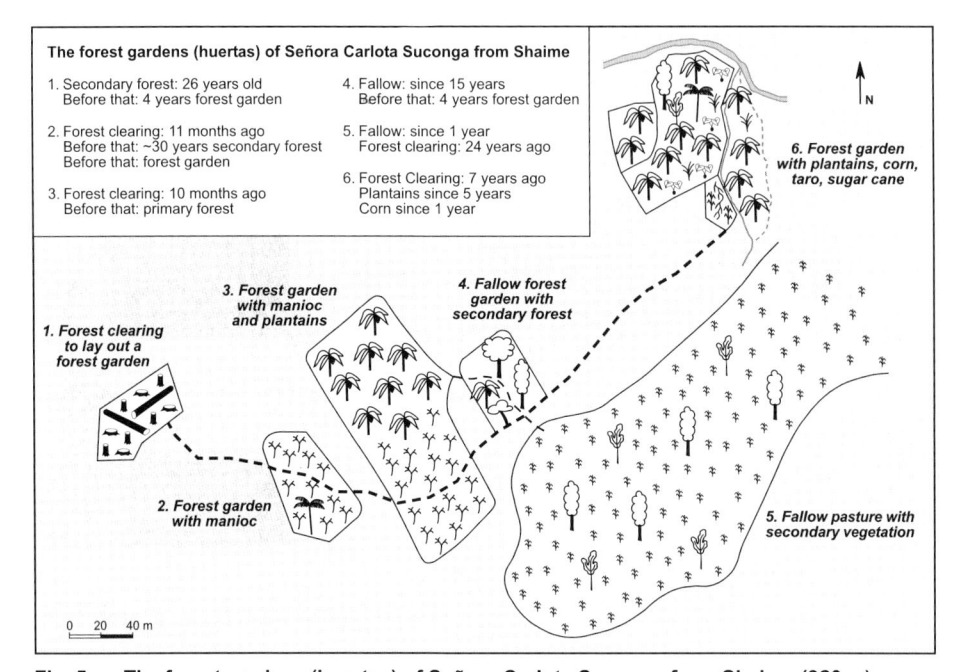

Fig. 5   The forest gardens (huertas) of Señora Carlota Suconga from Shaime (920 m)

Photo   P. Pohle

Photo 2   Richly chequered cultural landscape of the Saraguros on a steep slope of the Río Tibio valley with the scattered settlement of El Tibio (1,770 m)

Photo   P. Pohle

Photo 3   Deforested and overused agrarian landscape of the Mestizo-Colonos north of Loja

tivated plants. As it is commonly known, polycultures are more resistant to insect attacks and plant diseases.

- The natural fertilization of soils by mulching and the use of digging sticks and dibbles as a suitable form of cultivating the soil.
- Their sustainable use of a broad spectrum of wild plants in small quantities, satisfying only subsistence needs and avoiding over-harvesting.

If the Shuars' traditionally practiced and clearly sustainable plant biodiversity management is to be preserved, this is possible only by the legalization of their territorial claims and comprehensive protection of their territories; for example by demarcating reservations. This appears to be underway with the establishment of a so-called *Reserva Shuar* (NEILL 2005). Additionally, it is necessary to respect and support the Shuar's cultural identity, not only to avoid the loss of traditional environmental knowledge, in particular traditional plant lore. To improve livelihood in an economic sense, additional sources of financial income are essential. In this line the promotion of ecotourism, support of traditional handicrafts and the cultivation of useful plants for a regional market could be discussed.

While the Shuar's forest management can be evaluated as preserving plant diversity, the sustainability of the Saraguros' use of the environment has yet to be rated. Market-oriented stockbreeding has particularly led in recent decades to the rapid increase of pastures at the expense of forest. In spite of ecological conditions unfavorable to agricultural pursuits (steep V-shaped valleys, acidic soils, extremely high precipitation), these Andean mountain farmers have at least, by means of their intensive form of pasture management, succeeded in generating a sufficiently stable agrarian and cultural landscape (photo 2). In contrast to many completely deforested and ecologically devastated areas settled by Mestizo-Colonos (photo 3), the richly chequered agrarian landscape of the Saraguros presents, not only esthetically but also ecologically, a fundamentally more positive picture.

Among the Saraguros, initial attempts have also been elaborated to manage biodiversity in line with the concept "preservation through use". The first thing to be mentioned in this context is the keeping of home gardens with a wide spectrum of wild and cultivated plants, particularly woody species. With regard to the diversity of species, the remnants of forest still largely preserved in ecologically unfavourable locations are significant. In order to stem the further loss of biodiversity, however, it will be necessary to convince the Saraguros that in particular scrub- and wasteland (*matorral*) should be replanted with native tree species. The pressure on the tropical mountain forests caused by the pasturing economy will only be reduced, though, when the Saraguros can be shown a profitable alternative to it. As examples of promising endeavours in this context may be regarded:

- The selective timber production and replanting with native tree species as it is proposed by foresters (GÜNTER et al. 2004);
- The introduction of silvipastoral or agroforestry systems;
- The market-oriented gardening;
- The cultivation and marketing of useful plants, e. g. medicinal herbs;
- The promotion of "off-farm" employment opportunities;
- The payment for environmental services to protect the watershed area of Loja.

8 The Overexploitation of Natural Resources by the Mestizo-Colonos

The Mestizo-Colonos living today in the northern bufferzone of the Podocarpus National Park arrived from the 1960s onwards, encouraged by the national land reform of 1964. Most of the Mestizo-Colonos settling along the road between Loja and Zamora arrived only during the past 25 years. As colonizers they converted large areas of tropical mountain rainforests into almost treeless pastures (photo 3). To sustain livelihood they were forced to use even very steep and marginal areas for cattle ranching. Fire was the most common (easy and fast) way of clearing the forests to gain pasture land and as a consequence, large areas of forest were burned up, often in an uncontrolled way. The abandoned and scorched areas were immediately taken by the *llashipa* or bracken fern (*Pteridium aquilinium*). During the late 1970s and in the 1980s cattle ranching was displaced as main regional economic activity by timber extraction. This provoked the selective disappearance of many tree species with economic value and the alteration of adjacent forest areas, but as WUNDER (1996) pointed out, the need for new pasture land and not the extraction of timber was the principal reason for deforestation in this area.

At present, the economy of the Mestizo-Colonos is again based on cattle ranching, but the use of fire to open new pastures has declined, even though not disappeared. This has several reasons: First, many areas belong to the *Bosque Protector Corazón de Oro* or are part of the buffer zone of the Podocarpus National Park where it is forbidden to clear large forest areas and peasants face penalties if they do so. The road allows an easy control by the police and by the officers of the Ministry of the Environment in this area. Second, the remaining forests are today under private ownership. Therefore, uncontrolled fire can be a problem for peasants if it affects the neighbouring private properties. Since the remaining forest is nowadays further away from their housing and good quality timber is getting scarce in the forests, the Mestizo-Colonos are becoming more receptive to the idea of tolerating a variety of tree species in order to have shadow for cattle, timber for repairing fences, fuel, or material for construction. *Tabebuia chrysantha*, *Cedrela* spp., *Piptocoma discolor*, *Inga* spp. and *Psydium guayava* are among the most common tolerated species. Thus, the new pastures with tolerated tree species are easy to distinguish from elder ones, which have almost no growing trees on them.

Among the Mestizo-Colonos, the low profitability of the extensive form of cattle ranching has lead to a high in- and out-migration and a correspondingly high fluctuation in landownership and possession. To avoid poverty most of them have an alternative income, for example, from a small shop near the road, or a second occupation as day labourers. As a consequence, in marginal areas the pasture management is obviously neglected, mainly because of the limited availability of labour. Many Mestizo-Colonos are also supported by relatives who live in the city of Loja or by those who have emigrated to Cuenca or to Spain. Compared with the indigenous ethnic groups, the Mestizo-Colonos are much more heterogeneous. Extreme overexploitation is especially typical for newcomers and colonizers of the first generation mainly living in scattered *fincas* along the road between Loja and Zamora. In contrast, colonizers of the second or third generation living in village communities like in the upper Río Zamora valley have developed a more adapted and sustainable form of land use, in many respects similar to that of the Saraguros.

To avoid a further loss of biodiversity and to reduce the pressure on the tropical mountain rainforests caused by cattle ranching, profitable alternatives have to be offered. Similar measures as those presented above for the Saraguros are necessary: for example, timber production in the form of reforestation with native tree species, market-oriented gardening, cultivation of medicinal herbs, payment for environmental services. These measures should also take into account an improvement of the carrying capacity of existing pastures by introducing legumes (Leguminosae) and other fodder plants and improvement the cattle breeds and the veterinary services. In any case, measures to stop the loss of biodiversity in the area should take into account the difficult economic situation of most of the Mestizo-Colonos households. The prohibition of using fire and the establishment of protected areas have surely reduced the deforestation rate, but such measures do not face the real causes of deforestation and have increased the animosity against conservation.

9   Protecting Biological Diversity – From National Park to Biosphere Reserve

The Podocarpus National Park, covering a total of 146,280 ha, was established in 1982 as southern Ecuador's first conservation area, whose goal is to protect one of the country's last intact mountain rainforest ecosystems, one particularly rich in species and largely untouched by humans. The creation of a national park in the middle of a fairly densely populated mountain region necessarily gave rise to numerous conflicts of interest and use rights, for example, agrarian colonization, illegal timber extraction, conflicts about landownership and possession, mining activities, tourism.

The experience in international nature conservation during the past decades has shown that resource management, if it is to be sustainable, must serve the goals of both nature conservation and the use claims of the local population. The strategy is one of "protection by use" instead of "protection from use", a concept that has emerged throughout the tropics under the philosophy "use it or lose it" (JANZEN 1992, 1994). In the following, strategies are presented that show a way how people can benefit from the national park without degrading the area ecologically: the implementation of extractive reserves, the promotion of ecotourism and as the most promising approach the establishment of a Biosphere Reserve.

Given the high biodiversity of tropical rainforests and the fact that indigenous people in general have a comprehensive knowledge of forest plants and their utilization, it seems possible according to KRICHER (1997: 357) to view the rainforest as a renewable, sustainable resource from which various useful products can be extracted on a continuous basis. In view of the high number of plant species that are currently collected by extractivists in the surrounding of the Podocarpus National Park (e. g. the Shuar) the preservation of large areas of rainforest would make economic sense as well as serve the interests of conservation and preservation of biodiversity. Thus, the establishment of an extractive reserve could be suggested as an alternative to deforestation.

In line with the concept "protection by use" ecotourism can be structured such that it is compatible with conservation interests and serves the local economy as well. This is also the experience around Podocarpus National Park. The attraction of the park is clearly the tropical rainforest with its specific wildlife, particularly tropical birds, fewer visitors have botanical or eco-geographical interests. However, compared with other national parks in South America (e. g. the Manu National Park of Peru), southern Ecuador and the Podocarpus National Park are not a major tourist destination.

The most promising approach, in which conservational protection and sustainable development are the guiding principles, is the integrated concept of conservation and development exemplified by UNESCO's Biosphere Reserve (*UNESCO* 1984). The idea behind it is to mark out representative sections of the landscape composed of, on the one hand, natural ecosystems (core area) and, on the other, areas that bear the impress of human activity (buffer zone, development zone; Erdmann 1996). In Ecuador, three biosphere reserves have already been drawn up (*Ministerio del Ambiente* 2003). The establishment of such a reserve would also be desirable for southern Ecuador.[3] Alongside a strictly protected core area comprising Podocarpus National Park, it might encompass in a buffer zone cultural landscapes that have arisen historically (e. g. Vilcabamba) together with sanctuaries of indigenous communities (e. g. the proposed *Reserva Shuar*) and, in a development zone, areas of recent agrarian colonization.

Biosphere reserves are strongly rooted in cultural contexts and traditional ways of life, land use practices and local knowledge and know-how. In the buffer- and development zone of the Podocarpus National Park measures to be taken could rely on the rich ethno-specific traditions in forest- and land use practices by indigenous and local communities. In a first step it would be desirable to develop with the participation of the local communities' environmental management plans. On the one hand they could support the ethno-specific cultural tradition and strengthen the social identity of local communities. On the other hand these plans should comprise regulations for hunting, fishing, timber and plant harvesting, exclusion of human-created fires but also for house- and road construction. In southern Ecuador management plans are still in various phases of discussion and implementation. However, there is still a big gap between vision and reality. An intensive discussion was started about the participation of the population and of local NGOs about the assignment of environment competence to the regional-local administration. But, the realization of concepts like "cooperative management structure" could not consolidate and find political acknowledgement. According to Gallrapp (2005) it was failed to create effective platforms for participation and negotiation, to build up a common vision of the participating persons and to create a social awareness in order to implement new structures.

10 Perspectives

As the studies in southern Ecuador have shown, a realistic and sustained protection of nature in settled regions cannot be achieved without the participation of the local population, or without recognizing, protecting and promoting ethno-specific traditions of

---

[3] The proposal for the establishment of the Biosphere Reserve "Podocarpus – El Condor" has been approved by the UNESCO in September 2007.

preserving biodiversity. Thus the protection of biodiversity is intimately bound up with the protection and promotion of cultural diversity, particularly when, as in southern Ecuador, the "hot spots" of biodiversity coincide with those of cultural diversity. "Western" notions of nature conservation, as exemplified by national parks, as a rule pay insufficient attention to this cultural dimension. Thus concepts are needed that integrate the protection of both nature and culture, that is that practice both biological and cultural diversity. One approach along this line is the integrated concept of conservation and development exemplified by UNESCO's Biosphere Reserve.

Acknowledgements

We wish to thank all the inhabitants of the communities of Shaime, Chumbias, Napints, El Tibio and Sabanilla for their hospitality and generous participation in this study. We are grateful to the Herbario Reinaldo Espinosa of the Universidad Nacional de Loja (UNL) for their expertise and assistance with data collection and plant identification as well as for their logistical support. We would like to thank Eduardo Tapia for his expertise and assistance in collecting and processing geographical data. Our gratitude goes especially to the German Research Foundation (DFG) for supporting this research project.

References

BARTHLOTT, W.; W. LAUER; A. PLACKE; A. (1996): Global distribution of species diversity in vascular plants: towards a world map of phytodiversity. Erdkunde 50(4): 317–327.

BELOTE, J. D. (1998): Los Sarguros del Sur del Ecuador. Ediciones Abya-Yala, Quito.

BERKES, F. (1999): Sacred Ecology: Traditional Ecological Knowledge and Resource management. Philadelphia.

Die Erde (2001): Special issue ‚Tropische Wald-Ökosysteme'. Die Erde 132(1).

ELLENBERG, L. (1993): Naturschutz und Technische Zusammenarbeit. Geographische Rundschau 45(5): 290–300.

ERDMANN, K.-H. (1996): Der Beitrag der Biosphärenreservate zu Schutz, Pflege und Entwicklung von Natur- und Kulturlandschaften in Deutschland. In: KASTENHOLZ, H. G.; K.-H. ERDMANN; M. WOLFF (eds.): Nachhaltige Entwicklung. Zukunftschancen für Mensch und Umwelt. Berlin.

FAO (Food and Agriculture Organization of the United Nations) (2006): Global forest resources assessment 2005. Progress towards sustainable forest management. FAO Rome.

FUJISAKA, S.; G. ESCOBAR; E. J. VENEKLAAS (2000): Weedy fields and forests: interactions between land use and the composition of plant communities in the Peruvian Amazon. Agriculture, Ecosystems and Environment, 78: 175–186.

GALLRAPP, D. (2005): Naturschutz im Wandel? Vision und Realität integrativer Naturschutzkonzepte, am Beispiel des Podocarpus-Nationalparks in Südecuador. In: Tübinger Geographische Studien, 142: 417–441.

GERIQUE, A.; D. VEINTIMILLA (2007): Useful plants and weeds occuring in Shuar, Saraguro, and Mestizo communities. Checklist Reserva Biológica San Francisco (Prov. Zamora-Chinchipe, S. Ecuador). ECOTROPICAL MONOGRAPHS, 4: 237–256.

Gräf, M. (1990): Endogener und gelenkter Kulturwandel in ausgewählten indianischen Gemeinden des Hochlandes von Ecuador. München.

Günter, S.; B. Stimm; M. Weber (2004): Silvicultural contributions towards sustainable management and conservation of forest genetic resourcec in southern Ecuador. LYONIA, 6: 75–91.

Hamilton, L.; J. Juvik; F. Scatena (1995): The Puerto Rico tropical cloud forest symposium: Introduction and workshop synthesis. Ecological Studies 110: 1–19.

Janzen, D. H. (1992): A south-north perspective on science in the management, use and economic development of biodiversity. In: Sandlund, O. T.; K. Hindar; A. H. D. Brown (eds.): Conservation of Biodiversity for Sustainable Development: 27–52.

Janzen, D. H. (1994): Wildland biodiversity management in the tropics: where are we now and where are we going? Vida Silvestre Neotropical, 3: 3–15.

Kricher, J. (1997): A Neotropical Companion. Princeton.

Ministerio del Ambiente (2003): Ponencias del Ministerio del Ambiente para el Fortalecimiento y Consolidación del Sistema Nacional De Áreas Protegidas. Primer Congreso Nacional de Áreas Naturales Protegidas. Quito – Ecuador.

Müller-Böker, U. (1999): The Chitawan Tharus in Southern Nepal. An Ethnoecological Approach. Nepal Research Centre Publications, 21. Stuttgart.

Münzel, M. (1977): Jívaro-Indianer in Südamerika. Roter Faden zur Ausstellung, 4. Museum für Völkerkunde, Frankfurt/ M.

Münzel, M. (1987): Kulturökologie, Ethnoökologie und Ethnodesarrollo im Amazonasgebiet. Entwicklungsperspektiven 29. Kassel.

Münzel, M. (1989): Bemerkungen zum indianischen Umweltbewußtsein im Amazonasgebiet. Geographische Rundschau 41(7-8): 431–435.

Myers, N.; R. A. Mittermeier; C. G. Mittermeier; G. A. B. da Fonseca; J. Kent (2000): Biodiversity hotspots for conservation priorities. Nature 403: 853–858.

Nazarea, V. D. (ed.) (1999): Ethnoecology, situated knowledge/located lives. The University of Arizona Press, Arizona.

Neill, D. (2005): Cordillera del Cóndor. Botanical treasures between the Andes and the Amazon. Plant Talk, 41: 17–21.

Pohle, P. (2004): Erhaltung von Biodiversität in den Anden Südecuadors. Geographische Rundschau, 56(3): 14–21.

Pohle, P. (2008): The People Settled Around Podocarpus National Park. In: Beck, E.; J. Bendix; I. Kottke; F. Makeschin; R. Mosandl (eds.): Gradients in a Tropical Mountain Ecosystem of Ecuador. Ecological Studies, 198: 25–36. Berlin.

Posey, D. A. (1985): Indigenous management of tropical forest ecosystems: the case of the Kayapó indians of the Brazialian Amazon. Agroforestry Systems, 3: 139–158.

Posey, D. A.; W. Balée (eds.) (1989): Resource Management in Amazonia: Indigenous and Folk Strategies. Advances in Economic Botany, 7. New York.

Rudel, T. K.; W. Horowitz (1993): Tropical Deforestation, Small Farmers and Land Clearing in the Ecuadorian Amazon. Columbia University Press, New York.

Toledo, V. M.; B. Ortiz; S. Medellín-Morales (1994): Biodiversity islands in a sea of pasturelands: indigenous resource management in the humid tropics of Mexico. Etnoecologica 2(3): 37–50.

UNESCO (ed.) (1984): Action plan for biosphere reserves. Nature and Resources 20/4.

Van den Eynden, V. (2004): Use and management of edible non-crop plants in southern Ecuador. PhD Thesis in Applied Biological Sciences, Gent University, Belgium.

WARREN, D. M.; L. J. SLIKKERVEER; G. BROKENSHA (1995): The cultural dimension of development: Indigenous knowledge systems. London.
WATSON, J. W.; P. B. EYZAGUIRRE (2002): Home gardens and in situ conservation of plant genetic resources in farming systems. Proceedings of the Second International Home Gardens Workshop, 17–19 July 2001, Witzenhausen, Germany. IPGRI-Publications.
www.bergregenwald.de
WUNDER, S. (1996): Deforestation and the uses of wood in the Ecuadorian Andes. Mountain Research and Development, 16(4): 367–382.

Diversity, Complexity and Dynamics: Land Use Patterns in the Central Himalayas of Kumaon, Northern India

Marcus Nüsser[1] & Martin Gerwin[2]

Introduction

Natural resource use in high mountain areas is generally diverse as potentials, constraints and risks of agrarian practices are largely subjected to certain environmental conditions. The altitudinal zonation of climate and vegetation belts and the related ecological particularities stipulate vertical and seasonal mobility patterns to increase the availability and variety of natural resources. Consequently, land use systems have predominantly been interpreted in the context of local adaptation strategies within given spatial distribution and economic appraisal of natural resources. This proves especially true for peripheral mountain areas, where both land use and livelihood security primarily depend on subsistence farming, animal husbandry, and forest utilization (BISHOP 1990, STEVENS 1993, UHLIG 1995, NÜSSER 1998). Even if off-farm income opportunities have increased in almost every part of the South Asian high mountain rim comprising the Hindu Kush, Karakorum and Himalayas, agro-pastoral land use still constitutes the economic backbone of mountain communities. Beyond common features of combined mountain agriculture (EHLERS & KREUTZMANN 2000), regional agrarian practices vary between different sections of the high mountain belt of South Asia. Nonetheless, the analysis of regional land use cannot be reduced solely to aspects of adaptation to natural resource potentials and management strategies to mitigate the impacts of losses due to environmental hazards. A better understanding of land use change in both production patterns and livelihood strategies strongly depends on the integration of socioeconomic developments, cultural values, external influences, and the territorial dimensions of land tenure (KREUTZMANN 2004). Especially in the context of common property regimes and community-based institutions, the interaction of local norms, formal and informal regulations, and external development interventions all must be taken into consideration. Although integrated approaches are often postulated, changing strategies of resource use are predominantly conceived as a one-way street in which local actors react more or less uniformly to external impulses and restrictions.

The purpose of this paper is to analyze the diversity, complexity and dynamics of land use patterns in Kumaon, the north-eastern portion of the Indian state Uttarakhand. Much like in adjacent high mountain regions, agrarian land use in this part of the Central Himalayas takes place within a context of combined mountain agriculture across different altitudinal belts. The multiple territorial rights of access and natural resource utilization mainly results from ethnic segregation, settlement processes and external impacts. Most notably since the onset of British colonial rule in South Asia, the Central Himalayas have served as an effective barrier to the territorial expansion of diverse emperors. Nevertheless, the control over the vast natural resources of the moun-

---

[1] Ruprecht-Karls-Universität Heidelberg. Abteilung Geographie am Südasien-Institut, D-69120 Heidelberg; marcus.nuesser@uni-heidelberg.de.
[2] Ruprecht-Karls-Universität Heidelberg. Abteilung Geographie am Südasien-Institut, D-69120 Heidelberg; martin.gerwin@uni-heidelberg.de.

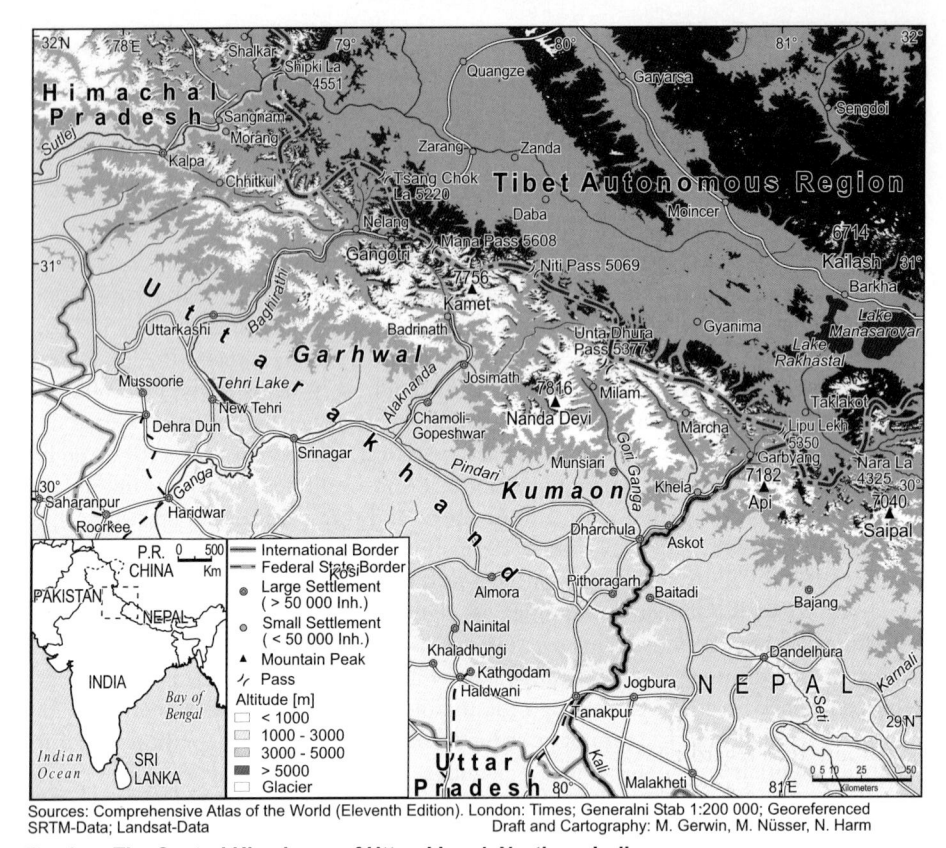

Fig. 1   The Central Himalayas of Uttarakhand, Northern India

tain belt, first and foremost forests, have always been an important matter of interest for various state agencies. In the given regional context of Kumaon, the ethno-linguistic group of the Bhotiyas deserves particular attention. Apart from their agro-pastoral land use patterns, the Bhotiyas practiced trade with nomads of the Tibetan Plateau until the war between India and China in 1962 and the resulting sealing of the border. The traditional land use system of the Bhotiyas can be illustrated using the case of the Gori Valley, which originates in the Trans-Himalayas and crosses the main ridge of the Greater Himalayas over a distance of roughly 100 km. Prominent massifs like Nanda Devi (7816 m), Hardeol (7151 m) and Panchchuli (6913 m) frame the upper portions of this transversal valley.

It remains to be seen how land-use patterns and strategies of resource use of the Bhotiyas have changed during shifts in political power and subsequent socioeconomic restructuring of the region. Whether or not or to what extent new strategies of land use have arisen or the existing ones have been modified remain open questions.

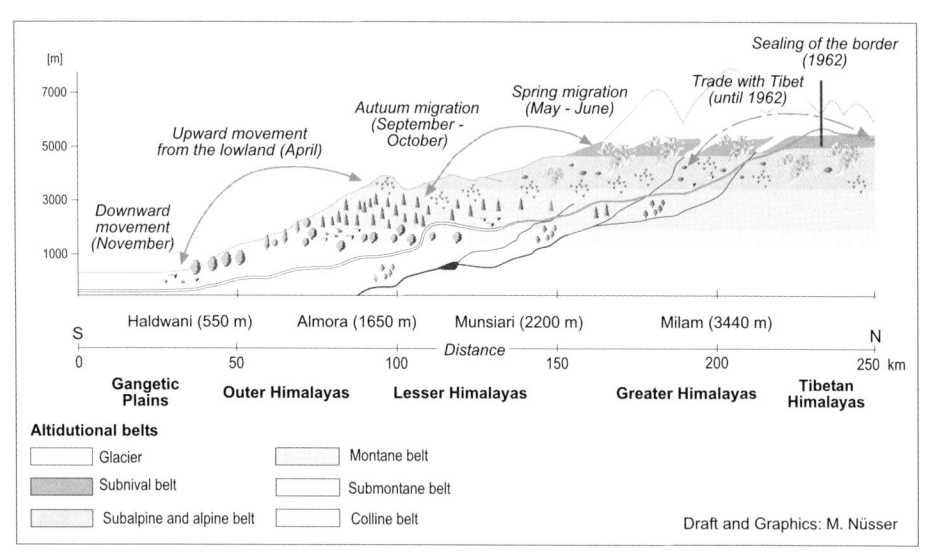

Fig. 2   Vertical Zonation and Traditional Mobility Patterns in Kumaon, Northern India

Environmental Zonation

The Kumaon and Garhwal Himalaya constitute the Indian federal state of Uttarakhand (fig. 1). Both regions span over several environmental zones with specific potentials and limitations regarding crop farming, forest and pasture use. The narrow belt of the Outer Himalayas arises from the northern parts of the Ganges Plains (*Terai*). Up to an altitude of approximately 1000 m this area is covered by tropical sub-humid Sal-forests (*Shorea robusta*), mostly reduced to steep slopes and almost exclusively protected as state-governed Reserved Forests. Without any visible disruption the range of the Lesser Himalayas, also known as *Himanchal*, follows in northern direction. This 70 to 100 km wide belt of mountains with an elevation of 1500 to 3000 m is dominated by *Pinus roxburghii*-forests. These conifers have been tapped for turpentine extraction since the early 20th century (AGRAWAL 2005). In the vicinity of settlements nearly the entire available land is terraced and cultivated for double cropping with rice, wheat, millet and different legumes.

The zone of the Greater Himalayas of Uttarakhand is about 30–50 km wide (Fig. 2). The scenery is dominated by glaciated mountain peaks, many of which exceed 6000 m and some 7000 m. The narrow transversal valleys are covered by montane forests and alpine grasslands, the former boasting a large variety of tree species. Evergreen oak forests, consisting of *Quercus semecarpifolia*, *Qu. dilatata* and *Qu. leucotrichophora*, alternate with areas dominated by conifers (*Abies spectabilis*, *Cupressus torulosa*) or deciduous trees, such as *Alnus nepalensis*, *Aesculus indica* or different species of maples (*Acer spp.*). The northernmost part of the transect is characterized by the Tibetan Himalayas. Located in the rain shadow of the main mountain range, the quantity of summer precipitation is rather low. During winter months however, the area is covered by thick and wet snow. Since the landforms have been shaped by glacial erosion processes, the valley

Photo 1 The transition zone between the montane and subalpine vegetation belts in the Gori Valley (M. Nüsser, October 2004)

bottoms are relatively wide and shallow. These localities are covered by alpine meadows and dwarf shrubs and are mainly used as pastures (photo 1).

Agro-Pastoral Resource Utilization and Trans-Himalayan Trade

The traditional system of agro-pastoral resource utilization of the Bhotiyas comprised pasturing, crop cultivation and forest use. Their seasonal transhumance linked all the aforementioned environmental zones to ensure an optimal use of all available fodder resources. Until the closure of the Indo-Tibetan border in 1962, they combined their agro-pastoral activities with Trans-Himalayan trade. Their winter migrations led them to the southern fringes of the mountain belt close to the Ganges Plain in search for valuable grazing grounds and trade markets (Hoon 1996). In this region, the towns of Ramnagar and Haldwani were already connected to the expanding railway system in the 1880s. During upward migration, the Bhotiyas drove their flocks to halting places (*Paraos*) in the montane belt in April and further up to the subalpine and alpine pastures between May and June. They returned to their halting places between September and October and reached the foothills again in November. This agro-pastoral migration cycle between summer and winter settlements was organized over a horizontal distance of roughly 200 km (fig. 2). Sheep and goats were the dominant animals for crossing the mountain passes to Tibetan trade markets during the summer months. Cattle and cross-breeds with yaks were suitable animals to deliver a sufficient amount of manure and to provide draught power (Traill 1832).

This system of agro-pastoral resource utilization is, however, subject to constant change, similar to other regions of the Central Himalayas. Summer migration to the *Bugyals*

Photo 2 Abandoned houses in Milam (3440 m), the uppermost settlement of the Gori Valley (M. Nüsser, October 2004)

for the purpose of high pasturing has reduced drastically. Apparently, due to the closure of markets for grain in Tibet, crop-farming in high altitude settlements is no longer profitable. During prosperous phases of trade crop-farming was conducted extensively and dominated by the growth of barley and buckwheat that were often irrigated due to the more arid conditions (PANT 1935). Presently, most of the buildings are abandoned and agricultural land is almost exclusively left fallow (photo 2).

The significant decrease in sheep and goats as well as the parallel increase of cattle (TRAILL 1832, *KBPF* 1947, Government of Uttaranchal 2003) indicate the transformation from transhumance towards sedentary agriculture. Nowadays cattle are the preferred species for agriculture in the middle sections of the High Himalayan valleys within the vicinity of the former trade depots. Since the montane forests often exhibit a herbaceous understorey, the forests also serve as important pastures for cattle, which is supplemented by stall-feeding during the winter months. This observation is reinforced by a considerable decline in the use of other transport animals, such as horses, ponies, and donkeys (*KBPF* 1947, Government of Uttaranchal 2003).

Downward migration to the lowlands and foothills is significantly reduced in present days. Only professional shepherds take remaining sheep and goats of the Bhotiyas down to these areas during the cold period. Due to legal regulations the shepherds are allowed to use one pasture for 10-15 days only. Contrarily, foreign livestock owners come to the valleys to graze their animals, mainly buffalos, between May and September.

Photo 3 Sacred birch forest above the village Martoli, Gori Valley (M. Nüsser, October 2004)

Socio-Cultural Landscapes

The mountain landscape of Kumaon and Garhwal is characterized by both its environmental and socio-cultural diversity (RAMAKRISHNAN 2003). Since ancient times the region has been regarded as the abode of gods and deities. Examples of the sacredness of landscape elements include the source of the Ganges at the Gangotri Glacier, the Nanda Devi as the seat of a goddess or the important Hindu temple at Badrinath. The spiritual meaning of mountain peaks, rivers or forests strongly correlates with regulations of natural resource use. Numerous sacred forest groves (photo 3) exhibit restrictions on the utilization of trees as fuel wood, building material or animal fodder. Traditional pilgrim routes not only lead to shrines of Uttarakhand, but connect the Himalayan valleys with the Tibetan Plateau, especially with Mount Kailash, the sacred mountain of Hindus and Buddhists.

The contact zone between the Hindu-dominated Indian Himalayas and the Buddhist-orientated Tibetan Plateau is an area where various cultural influences merge. The Bhotiyas, who live in this transition zone in Kumaon, Garhwal and in the northwestern parts of Nepal, are widely described as brokers (TRAILL 1832, HOON 1996). Their ecological knowledge and their ability to communicate in Tibetan languages are viewed as the main reasons for their monopoly in regional Trans-Himalayan trade (SRIVASTAVA 1966, BROWN 1984). Whereas, nowadays they characterize themselves as Hindus, former scholars and early explorers regularly emphasized their Tibetan origin and the Buddhist influences on their cultural practices to distinguish them from the major local Hindu population in these mountain valleys, the so-called Paharis (e. g. ATKINSON 1882). More recent research, however, alludes to the uncertainties of their geographical origin and, instead, emphasizes the heterogeneity within the group itself (SAKLANI 1998).

Development in a Contested Border Region: a Historical Perspective

A better understanding of regional land use patterns and development trajectories strongly depends on the integration of political processes and external influences in a historical perspective. Since the invasion of Kumaon and most parts of Garhwal by the Gorkhas in 1790, the area increasingly came to the fore of British strategic and economic interests (fig. 3). After the Gorkhas had been defeated by the British in 1815 and the first direct access to the Indo-Tibetan border was ensured (GILL 2000), the trade activities of the Bhotiya groups were reduced to the barter of locally needed products. Sugar, grain, and clothes from the Indian side were exchanged for salt, raw wool, animals, and borax from Tibet (PANT 1935, JOSHI & BROWN 1987). The more lucrative long distance trade of *pashmina* wool, a raw material for Kashmir shawls, was then exclusively conducted via Ladakh (LAMB 1986). After the invasion of Ladakh by the Dogras, the British colonial rulers initiated an eastward shift of trade to the effect that the routes monopolized by the Bhotiyas increased in significance (RIZVI 1999). Towards the end of the nineteenth century the reduction of taxes and transit duties by the British as well as the onset of mechanized wool production in the Indian plains pushed the Bhotiya's trade volume to its highest rate (ATKINSON 1882, SHERRING 1906). The summer settlements in the subalpine and alpine belts were prospering places during this period of time. From the 1920s on, however, trade patterns in general and wool-related trade in particular declined, the main cause being construction of roads connecting the existing railheads with the Lesser Himalayas. These infrastructural developments have been associated with increasing imports of cheaper wool products from Europe and (sea-)salt from the Indian lowlands (WALTON 1910, FÜRER-HAIMENDORF 1975). Additionally, due to the discovery of new natural deposits of borax in other parts of the world as

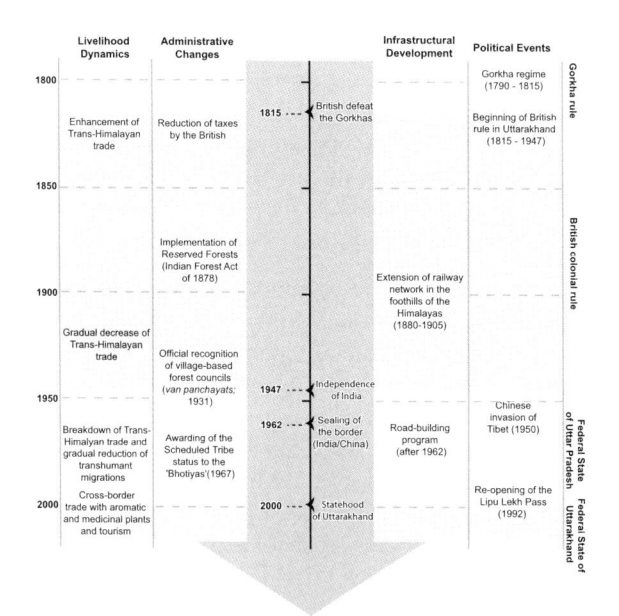

Fig. 3   Historical Events and Developments in Kumaon, Northern India

well as the availability of artificial substitutions of this mineral, the products traded and manufactured by the Bhotiyas became less competitive (SPENGEN 2000). The most important factor leading to a significant reduction and finally to the complete breakdown of Trans-Himalayan trade, however, was the Chinese invasion of Tibet in 1950 and the Indo-China war in 1962 (SRIVASTAVA 1966).

After the border war between India and China both nations continued to expand their presence in the contested border region. One of the crucial forms of intervention by the Government of India was a large road construction program, which has opened up the lower parts of most mountain valleys stretching towards Tibet for military purposes (RAWAT & SHARMA 1997). High altitude villages along the former routes were progressively abandoned as a result of the collapse of Trans-Himalayan trade and increased road accessibility of the main villages. The example of the village of Milam in the upper Gori Valley (fig. 1) exemplifies this development. Whereas a figure of 1,500 inhabitants was given for the early 19th century (TRAILL 1832) and 600 families were counted in the 1930s (PANT 1935), only 300 families were left in 1950 (Murray 1951) and the number was further reduced to only 18 families living in the village in 2004 (NÜSSER 2006). The former winter settlement and trade depot of Munsiari gradually became the central municipality and administrative centre of the Gori Valley in the course of the 20th century. Infrastructural investments and funds allocated by the Indian Government, such as for the construction of school or the public administration, have further promoted this growing significance with numerous non-farm employments. The Bhotiyas no longer constitute the majority in the Gori Valley. Instead people who are administratively counted as Paharis have formed the majority of the population in more recent times (Government of India 2003). It is estimated that about three quarters of the Bhotiyas have re-settled in other areas of Uttarakhand or the Indian plains due to the changes in the valley (PANGTEY, oral information). However, the shops at the market centre of Munsiari are almost exclusively in the hand of influential Bhotiya groups, who also dominate local politics. Quite clearly, these groups have been able to use their profits, prestige and education gained from trade and the associated dynamics of the borderland, but a further crucial point marks the awarding of the scheduled tribe status in 1967 for which the identification of distinguishing socio-cultural features is vital (BERGMANN et al. 2008). It was this status that guarantees them access to employment in government service, admission to universities, and reserved seats in the legislative bodies.

Additionally, new income sources have been opened up including the rise of tourism enterprises. The support and expansion of tourism to generate income is one strategy of the newly formed government of the hill state of Uttarakhand. Since this state was carved out of the most populated Indian state of Uttar Pradesh in 2000, the appointed government promised to take the specificities of the mountain environments and its inhabitants seriously (RANGAN 2004). Local inhabitants may apply for credits to set up tourism projects, and especially in and around Munsiari new enterprises, such as hotels, offices and homestay arrangements have been established. However, the allocating of funding and the realization of business are matters of local politics and numerous disputes arise on this topic within the area. It is also anticipated to revitalize activities in the upper part of the valley such as mountaineering and trekking, which in turn require new forms of infrastructural investment and the regulation of natural resources.

Politics of Natural Resource Regulation in Historical Dimension

Environmental regulations and resource protection in Kumaon have a long history and are deeply interwoven with changes in land use patterns. Colonial and postcolonial measures of resource control intermingle with local land use patterns and livelihoods. From the mid-nineteenth century onward, the main focus of British politics in Kumaon shifted from the motives of trade control and geopolitical influence towards the extraction of biological resources. During this period administrative structures were implemented and strengthened in order to increase taxes from agricultural land, forests and livestock (ATKINSON 1882; Political Department 1895). Apart from demarcations of government land and village boundaries for revenue purposes, numerous restrictions were imposed on the use of forests and grasslands. These regulations started in the foothills and Lesser Himalayas, where sal and pine forests had been protected for the generation of railway sleepers (AGRAWAL 2005). These interventions directly interfered with the traditional transhumance patterns of the Bhotiyas. Examples include the assignment of certain market places, halting stages and winter habitations during the annual migration cycle (*Forest Department* 1905). After the British administration had extended its environmental interventions into the northern parts of Kumaon at the end of 19th century, most of the forest and grass areas were bordered either as Reserved Forests or as Civil Forests. In both cases strict rules were imposed on the local population and territorial rights of access and utilization were demarcated and assigned village-wise.

The constitution of *van panchayats* as village-based administrative bodies for the regulation of forest use in the year 1931 marks a milestone in the history of natural resource management in Kumaon, following massive protests of local groups against the repressive colonial interventions. These decentralized councils are awarded a certain amount of freedom regarding the implementation of local rules for grazing, fuel wood and leaf litter collection in the areas assigned to them. As a matter of fact, these institutions represent one of the oldest surviving examples of formally approved agreements between state authorities and local forest users (AGRAWAL & OSTROM 2001). Presently, there are around 12,500 of these councils in the entire Kumaon (AGRAWAL 2005). Although this system of resource control dates back to the time of colonial rule, most *van panchayats* were established after independence. More recently, the control over medicinal plant collection (photo 4) arises as a further challenge for these institutions.

The Uttar Pradesh Zamindari Abolition and Land Reforms Act of 1950, which was drafted with the intention of transferring agricultural land to the actual cultivator, is considered to be one of the far-reaching forms of government intervention in agrarian resource use (RAHUL 1968, Government of Uttar Pradesh 1967). As a consequence of this land reform the Bhotiyas have almost completely lost their access to pastures and land in the foothills and the Lesser Himalayas and have therefore been forced to reduce their seasonal migrations (NAUTIYAL et al. 2003). The boundaries of the revenue villages were re-drawn and a new land and forest settlement has been carried out in the 1960s, also affecting the *van panchayat* areas (Government of Uttar Pradesh 1967). While the settlement act was designed to correct the colonial regulations, it was also influenced by local power relations. The result of this latest settlement, subsequent re-negotiations

Photo 4   Medicinal plant collection in the Ralam Valley (M. Nüsser, August 2005)

and re-allocations is a current patchwork of unequally sized village forests. In fact, even some villages are completely left out and are without any officially approved access to forests and grassland areas. The most striking example of size variations of village forests can be found in the alpine parts of the valley and the spatial arrangements around Munsiari in the Gori Valley. While some villages are almost completely abandoned in the upper portions, these resource user groups exhibit by far the largest *van panchayats* with vast alpine pastures. The area of the *van panchayat* of Milam for instance encompasses 35,081 ha of land. The areas around the former trade depot Munsiari by contrast are characterized by an intricately nested and interwoven mix of regulations. Here the *van panchayats* measures just 20–100 ha (Foundation for Ecological Security 2003).

In Munsiari people allude to an insufficient access to forest and grassland resources as one reason for the reduced number of livestock. Since some of the forest councils rely on paid guards or control their areas on a rotational base, people fall back on Reserved Forests that are situated nearby. In these forests resource control is often less strict or even non-existent. The limited access to the essential resources of mountain livelihood is also seen as one reason for the summer migrations to the upper valley zones by at least a fraction of the families. In these upper parts of the valley the access to grazing grounds is less restricted or people are members of a *van panchayat*. Forest products are regularly auctioned and not all persons are able to participate equally. Quite often non-farm income decides upon supply. In addition to the officially sanctioned rules, informal agreements exist. Beside ethnic affiliation and social position, customary rights permit access to forests and grasslands.

Conclusions

In spite of the various factors of transformation, the seasonally and spatially highly diversified 'combined mountain agriculture' still plays a vital role for a large number of households in Kumaon. Land use patterns and livelihood strategies shall be differentiated not only as for environmental issues, but according to such factors as ethnic and socio-cultural affiliation and territorial rights of resource access. A better understanding of the diversity, complexity, and dynamics of regional land use systems requires identifying and evaluating the historical dimension of human-environmental interaction. Changing land use structures and functions reflect the supra-regional political and socioeconomic developments. Land use change, however, has evolved rather slowly and gradually. For instance, with regard to common property regimes and community-based institutions colonial and postcolonial forms of intervention are clearly visible in present day land use patterns and mountain livelihoods.

The case of the Bhotiyas demonstrates their flexible response to changing political, economical and socio-cultural constellations in the borderlands of Kumaon. This includes the tangled relationship between strategies of agrarian subsistence production and non-agricultural employments. After the abandonment of the Trans-Himalayan trade, the upper valley sections have been transformed from transit corridors to new peripheries. This functional change has been reflected in reduced land use intensity accompanied by new income options such as collection and cultivation of medicinal and aromatic plants. The administrative and economic centres exhibit a heterogeneous mix of livelihoods, institutions and resource use strategies. New forms of income such as mountain tourism or shop-keeping have arisen, accompanied by combined mountain agriculture in the vicinity of the permanent settlements. The latest development of trade relations, however, is characterized by the re-opening of the Lipu Lekh Pass in the northeastern corner of Kumaon.

References

AGRAWAL, A. (2005): Environmentality. Technologies of Government and the Making of Subjects. Durham.

AGRAWAL, A; E. OSTROM (2001): Collective Action, Property Rights, and Decentralization in Resource Use in India and Nepal. Politics & Society 29: 485–514.

ATKINSON, E. T. (1882): The Himalayan Gazetteer. Volume Two, Part One. (Originally published under the title "The Himalayan Districts of the North Western Provinces of India"). Reprint 1996, Dehra Dun.

BERGMANN, C.; M. GERWIN; M. NÜSSER; W. S. SAX (2008): Living in a High Mountain Border Region: The Case of the 'Bhotiyas' of Uttarakhand, India. Submitted to Journal of Mountain Science.

BISHOP, B. (1990): Karnali under Stress: Livelihood Strategies and Seasonal Rhythms in a Changing Nepal Himalaya. Geography Research Paper 228–229, Chicago.

BROWN, C. W. (1984): The Goat is Mine, the Load is Yours: Morphogenesis of "Bhotiya-Shauka", U.P., India. Lund Studies in Social Anthropology 1, Lund.

EHLERS, E.; H. KREUTZMANN (2000): High Mountain Ecology and Economy: Potential and Constraints. In: EHLERS, E.; H. KREUTZMANN (eds.): High Mountain Pastoralism in Northern Pakistan. Erdkundliches Wissen 132, Stuttgart, 9–36.

Forest Department (1905): Petition Received from Certain Bhotias Complaining of the Insufficiency of Grazing near Ramnagar, dated 25th April 1905. File 61/1903 – 1907, Commissioner Office Records, Box 147. Regional Archives, Naini Tal.

Foundation for Ecological Security (2003): A Biodiversity Log and Strategy Input Document for the Gori Valley. Munsiari.

Fürer-Haimendorf, C. von (1975): Himalayan Traders: Life in Highland Nepal. London.

Gill, B. G. (2000): India's Trade with Tibet: Early British Attempts. The Tibet Journal 25: 78–82.

Government of India (2003): Census of India 2001. District Pithoragarh, Uttaranchal. Office of the Registrar General and Census Commissioner. New Delhi.

Government of Uttar Pradesh (1967): Uttar Pradesh Gazetteer – Pithoragarh District. Pithoragarh.

Government of Uttaranchal (2003): Livestock Survey for the Tehsil Munsiari, 2001. Veterinary Department of the District Pithoragarh. Pithoragarh (unpublished).

Hoon, V. (1996): Living on the Move: Bhotiyas of the Kumaon Himalaya. New Delhi.

Joshi, M. P.; C. W. Brown (1987): Some Dynamics of Indo-Tibetan Trade through Uttarakhanda (Kumaon-Garhwal), India. Journal of the Economic and Social History of the Orient 30: 303–317.

KBPF (1947): The Report of the Kumaon Bhotiya Peoples' Federation. Submitted Before the Minority Sub-Committee of the Constituent Assembly's Advisory Committee on 17th April 1947. Kindly handed over by Dr. S. S. Pangtey (Sep. 2006).

Kreutzmann, H. (2004): Pastoral Practices and their Transformation in the North-Western Karakoram. Nomadic Peoples 8: 54–88.

Lamb, A. (1986): British India and Tibet, 1766 – 1910. London.

Murray, W. H. (1951): Scottish Himalayan Expedition, 1950. London.

Nautiyal, S.; Rao, K. S.; Maikhuri, R. K.; K. G. Saxena (2003): Transhumant Pastoralism in the Nanda Devi Biosphere Reserve, India. A Case Study in the Buffer Zone. Mountain Research and Development 23: 255–262.

Nüsser, M. (1998): Nanga Parbat (NW-Himalaya): Naturräumliche Ressourcenausstattung und humanökologisches Gefügemuster der Landnutzung. Bonner Geographische Abhandlungen 97, Bonn.

Nüsser, M. (2006): Ressourcennutzung und nachhaltige Entwicklung im Kumaon-Himalaya (Indien). Geographische Rundschau 58: 14–22.

Pant, S. D. (1935): The Social Economy of the Himalayans. Based on a Survey in the Kumaon Himalayas. London.

Political Department (1895): Correspondence of J. V. Stuart, Esq., Deputy Commissionar Kumaon of Almora to Commissioner, Kumaon Division, dated 31th January 1895 (Proceedings for April 1895 Nos. 27–33 & 36). File 24/1896 – 1897, Commissioner Office Records, Box 50. Regional Archives, Naini Tal.

Rahul, R. (1968): The Development Program in the Himalaya. Asian Survey 8: 666–677.

Ramakrishnan, P. S. (2003): The Sacred Ganga River-based Cultural Landscape. Museum International 55: 7–17.

Rangan, H. (2004): Form Chipko to Uttaranchal: The Environment of Protest and Development in the Indian Himalaya. In: Peet, R.; M. Watts (eds.): Liberation Ecologies: Environment, Development and Social Movements (second edition). London, 371–393.

Rawat, D. S.; S. Sharma (1997): The Development of a Road Network and Its Impact on the Growth of Infrastructure: A Study of Almora District in the Central Himalaya. Mountain Research and Development 17: 117–126.

Rizvi, J. (1999): Trans-Himalayan Caravans. Merchant Princes and Peasant Traders in Ladakh. New Delhi.

Saklani, D. P. (1998): Ancient Communities of the Himalaya. New Delhi.

Sherring, C. A. (1906): Western Tibet and the British Borderland. The Sacred Country of Hindus and Buddhists. With an Account of the Government, Religion and Customs of its Peoples. London.

Spengen, W. van (2000): Tibetan Border Worlds. A Geohistorical Analysis of Trade and Traders. London.

Srivastava, R. P. (1966): Tribe-Caste Mobility in India and the Case of Kumaon Bhotias. In: Fürer-Haimendorf, C. von (ed.): Caste and Kin in Nepal, India and Ceylon. Anthropological Studies in Hindu-Buddhist Contact Zones. New York, 161–212.

Stevens, S. F. (1993): Claiming the High Ground. Sherpas, Subsistence, and Environmental Change in the Highest Himalaya. Berkeley, USA.

Traill, G. W. (1832): Statistical Report on the Bhotiya Mehals of Kumaon. Reprinted in: Joshi, M. P.; A. C Fanger; C. W. Brown (eds.) [1991 – 1992]: Himalaya: Past and Present, Vol. II. Almora, 99–154.

Uhlig, H., edited by H. Kreutzmann (1995): Persistence and Change in High Mountain Agricultural Systems. Mountain Research and Development 15: 199–212.

Walton, H. G. (1910): British Garhwal. A Gazetteer, Volume XXXVI of the District Gazetteers of the United Provinces of Agra and Oudh. Reprint 1989, Dehra Dun.

Diversity in Mountain Tourism: the Example of Kyrgyzstan

Andrea Schneider[1] & Jörg Stadelbauer[2]

A Periphery of Global Tourism

Tourism is today one of the leading sectors of the world economy and accounts for, depending on the definition, 10.4 to 12 percent of worldwide GNP. Particularly mountain countries, which have whosescenic resources that offer prime economic potential, are dependent upon tourism development: As a pioneer, Nepal has developed a system of trekking tours with regional centres. With a strict limitation of visitors, Bhutan focuses on more affluent tourists. Among the developed countries, Switzerland and Austria obtain a considerable amount of their GDP from mountain tourism. In other countries, such as Morocco, New Zealand, the western part of the USA and Canada, France, Spain, Slovenia, Bulgaria, Japan etc., mountain tourism is an essential sector of the economy. Kyrgyzstan, as a successor state of the Soviet Union, is a newcomer in the global tourism industry. However, even in Soviet times, there was a fledgling mountain tourism industry.

The leading idea of diversity in tourism is present in Kyrgyzstan in two different ways: on one hand, successful tourism needs a certain diversity of offers, on the other hand, the multiple combinations of offers, entrepreneurial organisations, service chains, target groups and travel incentives lead to various spatial-/agent-systems. When discussing this, the particularities of Kyrgyzstan's Soviet heritage have to be taken into account as well as the specific cultural features of the country's traditional nomadic society.

Therefore, the central questions for this article are as follows:
- Which Soviet heritage and which spatial structures in tourism did Kyrgyzstan take over after independence?
- Which concepts for the development of tourism did Kyrgyzstan adopt from other mountain countries?
- Which spatial structures and which consequences on other areas of life resulted from the latest development of tourism in Kyrgyzstan?
- Which stakeholders and other agents are involved in tourism in Kyrgyzstan and what spatial patterns of actions do they form?

Which inter- and intra-systemic relationships exist in the various tourism segments?

The objective is to understand answers to these questions as elements of a socio-economic system and furthermore to link them to other systematic correlations in mountainous regions. The analysis aims at identifying the spatial-agent systems with a broad concept of agents. Thus, economic stakeholders as well as participants on both the demand and supply sides are considered as agents.

---

[1] anschneider@web.de
[2] University of Freiburg, Department of Human Geography, D-79085 Freiburg; joerg.stadelbauer@geographie.uni-freiburg.de.

The Framework for Tourism in Kyrgyzstan: The Search for a Share in the Global Market

Tourism is a major economic force in the world. In 2007, nearly 900 million international tourism arrivals were registered worldwide, the total number of arrivals (including multiple counts) was 5.6 billion for forms of tourism and other travel reasons. The global value added chain of tourism is estimated to be US $4.1 trillion of which, however, Central Asia only has a very small share. According to the World Tourism Organization (UNWTO), Kazakhstan, Kyrgyzstan and Uzbekistan had a total share of 3.75 million international tourism arrivals in 2004. Of these, 84 percent were accounted for by Kazakhstan – among them many visitors of the Russian, Ukrainian and German diaspora with relatives in North Kazakhstan. Even if the unofficial small numbers of tourists in Tajikistan and Turkmenistan are added to the numbers of the UNWTO statistics, the five Central Asian states have a share of less than one per thousand of global tourism. With a turnover of only US$ 811 million, their financial share only accounts for 0.2 per thousand. Therefore, Central Asia can be considered to be on the periphery of the global tourism industry.

In Kyrgyzstan, tourism has a share of 3.6 percent of GDP according to national statistics. On average, 6,500 people are employed in tourism each year. In 2006, 765,850 international tourists visited Kyrgyzstan, 62.2 percent of these were from Kazakhstan and 12.4 percent from Uzbekistan. These numbers comprise all travel undertaken for all motives including business trips. A total of 257,000 overnight stays were registered in 123 hotels, which equates to an occupancy rate of only 7 percent (*Kyrgyzstan v cifrach 2007*, 2007: 290 sqq.). Despite its minor importance for the national economy, tourism in Kyrgyzstan is considered to have potential for further development. The IMF (International Monetary Fund) expects that ca. 2 million international tourists could create an income of US$ 400-500 million in Kyrgyzstan annually. To reach these numbers however, investment of about US$ 100 million would be necessary in the time period from 2007 to 2010 (*IMF 2007*: 38).

For 2004, the spatial structure of some basic data about tourism in Kyrgyzstan can be presented in more detail (Fig. 1 & 2), since a comprehensive census was carried out on 1 August 2004 (*Uslugi…*, 2005). With nearly 400,000 international arrivals, a respectable number of tourists visited the country in 2004, whereas in 2005, a decline of about 20 percent was recorded. This can be explained by the political insecurity in Kyrgyzstan after the forced resignation of President Akaev, which particularly led to a decrease in visitors from Kazakhstan (*Tourism v Kyrgyzstane*, 2006: 27).

Historically, mountain tourism was already important in Soviet times. However, its focus was in the Caucasus rather than in the Tien-Shan of Central Asia (STADELBAUER 1983). Nevertheless, the soviet tourism organisation also tried to develop facilities for the Central Asian mountain region and for the Altai as well as to some extent even for some south Siberian mountain ranges and the Carpathians (PREOBRAZHENSZKY & KRIVOSHEYEV 1980).

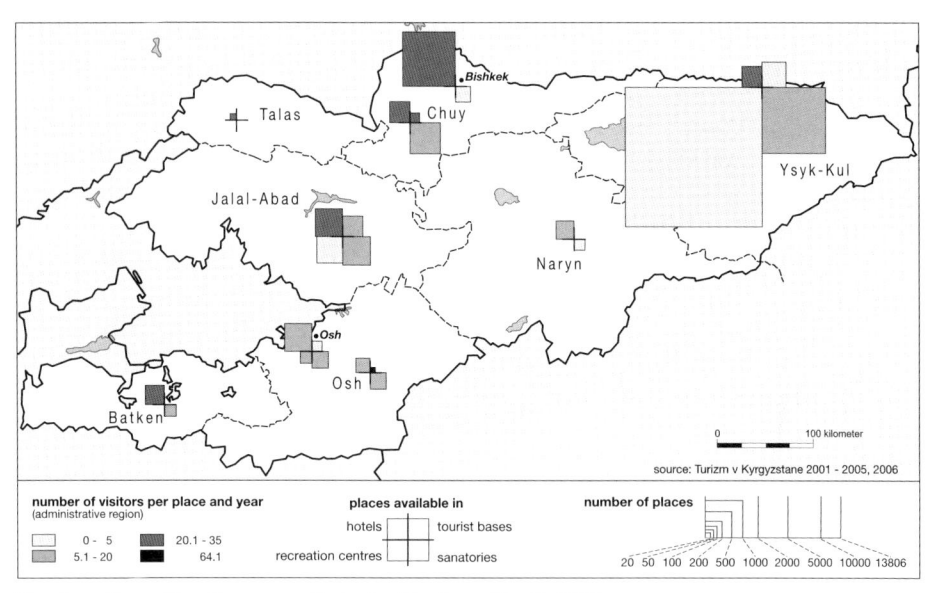

Fig. 1   Use of hosting capacity in the Kyghyz oblasti, 2005

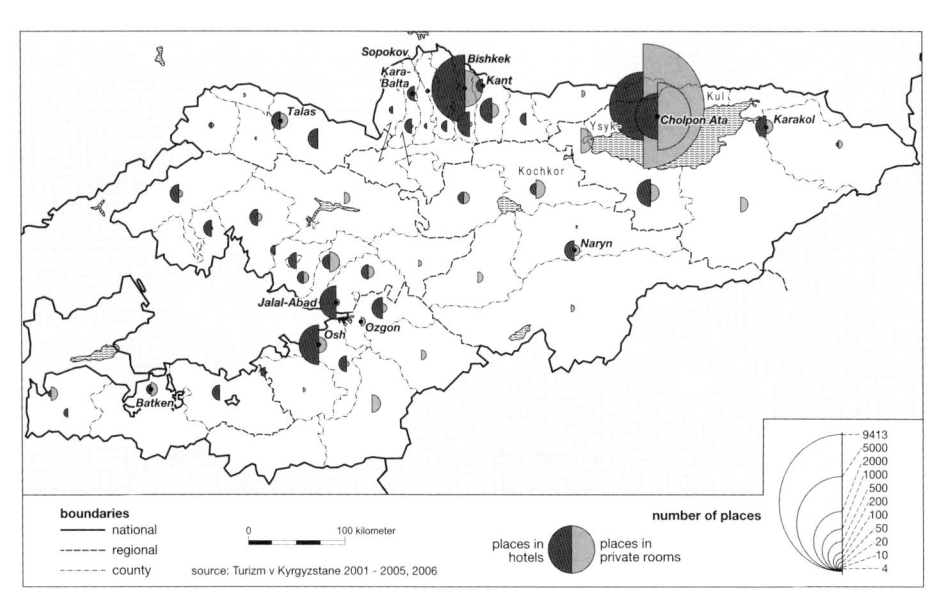

Fig. 2   Hosting capacity in hotels and private homes, 2005

Kyrgyzstan: Potential for Tourism in an Outstanding Natural Setting

In a nation-wide regionalisation, 14 resorts and recreational zones and 10 mountainous and alpinist zones were identified in Kyrgyzstan. For cultural tourism, however, Kyrgyzstan only has a very small potential and cannot keep up with its neighbour Uzbekistan in this market segment. Kyrgyzstan's capital Bishkek, as well as Karakol, are relatively young cities and their history can be largely attributed to the Russian colonial time since the 19th century. Few historical buildings still exist in Kyrgyzstan – among them buildings in Osh and Uzgen, the Burana Tower, the citadel of Koshoy Korgon and the restored caravanserai Tash Rabat. They can all be regarded more as individual objects along the way than as primary destinations of a tour. Therefore, Great Silk Road tourism in Kyrgyzstan can only be successful in cooperation with other countries that are located on the historical road network.

Concerning the scenic resources, the situation is different: Many peaks of the Tien Shan range are accessible for alpinism. Particularly the Zailiyskiy Alatoo and Kungei Alatoo at the Kyrgyz-Kazakh border have been made accessible for trekking tourism since the 1980s. The Terskey Alatoo, the Talas Alatoo and the Ferghana range, offer further trekking and mountaineering potentials. Alpinism and trekking demand a relatively high effort in planning and transportation as further development of supporting infrastructure such as accommodation and rest stops are required. However, some progress has been made and further measures have been taken conceptually as well as practically.[3] Thus, the rich potential of relatively unspoiled nature is probably the most important potential for attracting international tourism to Kyrgyzstan. For trekking tourism, it is intended to also attract the young domestic tourists and visitors from other countries of the CIS. With this aim, the Trekking Union of Kyrgyzstan was founded in 2005 in cooperation with tourism experts from Norway.

The shore area of Ysyk-köl is relatively well developed for tourism as research and planning for recreational and mountain tourism started in the 1960s/1970s. With the organisation of tourism under Soviet conditions it was hoped to attract a high number of Soviet visitors (PREOBRAZHENSKY & KRIVOSHEYEV 1980: 160ff.). Although many sanatoria, health resorts and rest homes had to close after Kyrgyzstan's independence, recently, their capacity is, at least partly, being reused due to an increase in demand, predominantly from Kazakhs (ASYKULOV & SCHMIDT 2005).[4] As a whole, 150 facilities for organised tourism were registered in Kyrgyzstan in 2005. Resort and recreational tourism, which is concentrated mainly on the Ysyk-köl region, accounts for 90 percent of the tourism revenues of the country.

Types of Tourism in Kyrgyzstan

At least four major types of tourism with different levels of participation can be distinguished in Kyrgyzstan, leaving out business trips, which concentrate mainly on Bishkek, cross-national transactions of Chinese tradesmen and various facets of political travels including consultants:

- Alpinist and mountaineering tourism is concentrated on high mountain tours and peaks. As it leads into uninhabitable and inhospitable areas, it needs to be supported by extensive logistical infrastructure. Detailed research work on this type of tourism does not exist.

[3] A recently completed diploma thesis in the field of cartography deals with the development of a long distance trail in the Ysyk-köl region (HAHMANN 2007).
[4] In an attempt to transform the whole region of the Ysyk-köl into a biosphere reserve, a handbook for the development of tourism (*GTZ* 2002) and an inventory of cultural assets (UHLEMANN 2003) were published.

- Community-based tourism (CBT) and trekking tourism lead into rural areas, are connected to existing settlements and can be seen as being similar to the traditional altitudinal zones of mountain nomadism as it includes settlements in the valleys as well as high altitude summer pastures (*jailoo*). In the past years, CBT has developed into a system focusing on regional development in rural areas.
- Recreational tourism in the Ysyk-köl region: The Soviet development of tourism in this region has been continued in a slightly modified way in post-Soviet times and is based on a network of recreational homes, sanatoria and other facilities in Ysyk-köl. It involves the local population as they offer accommodation and other services.
- Relicts from Soviet recreational tourism in other areas, i. e. in Arslanbob. They are characterised by massive problems of adaptation if none of the models and types of tourisms mentioned above can be adopted.[5]

Alpinism in Kyrgyzstan

Historical Review

The tradition of high mountain tourism in Kyrgyzstan dates back to the explorations of Pjotr Semyonov Tyan-Shanskiy (access to Sary-Djaz glacier in 1856/57 and description of the peak Khan-Tengri) as well as the descriptions of Gottfried Merzbacher (1902/03). In 1927, the first group for mountain tourism was founded in Karakol. In 1937, the Mountaineering Club was founded as a coordinated group of Soviet alpinists in the State Committee for Physical Culture and Sports so that it became possible to train the first alpinists from Kyrgyzstan. In the period of war, Kyrgyzstan also opened up for alpinists from other regions. Today, alpinism in Kyrgyzstan is controlled by the State, but organised by private tour operators who sub-contract work in order to provide all necessary services, such as obtaining the required permits.

Recent Destinations and Offers

Peaks which have been made accessible for alpinists are listed in various offers. However, not all peaks can be reached, since some of them are located close to the borders. Two examples shall be examined here to display available offers, organisation and spatial linkages[6].

The highest mountain in Kyrgyzstan is Peak Pobeda (7439 m). The best time to ascend is considered to be between 25 July and 25 August. The ascent to Peak Pobeda is comparable in terms of technical difficulty to the ascent of peaks in the Himalaya. The most popular route follows the relatively safe central slope to Pobeda West Peak via the Dikiy pass.

Peak Lenin (7134 m) can be reached from the north without specific training, particularly in the time between 20 June and 21 August. Peak Korzhenevski (7105 m) and

---

[5] This aspect was displayed by Kirchmayer 2005 (cf. Kirchmayer & Schmidt 2005) and shall not be examined here in further detail.
[6] The following summary refers to information on the websites of the Kyrgyz Alpine Club and of some tour organiszers; cf. www.kac.centralasia.kg/mountaineering%20possibilities%20kp.htm; www.tien-shan.com; www.Dtmc.centralasia.kg <20.04.2008>.

Peak Ismail Somoni (formerly Peak Kommunizm – which, at 7485 m, is the highest peak of the former Soviet Union) are located at the border of Kyrgyzstan and Tajikistan and are accessible from the same base camp. Usually, an expedition to the latter two lasts four weeks. It starts with the journey from Manas Airport to Dushanbe, followed by transfer by helicopter to the base camp at Moskvina glacier and a time of acclimatisation at Onfortambek glacier at 3600 m. After 6 days, the Korzhenevski Camp in 5100 m is reached. From here, Vorobyov Peak (5685 m) or Chetyrekh Peak (6230 m) are climbed first and after a day of rest at the base camp, the ascent to Korzhenevski Peak, which lasts 5 days, follows. After another day of rest, the ascent to Piak Ismail Somoni starts. This takes another six to seven days and forms the last part of the expedition before the departure to Bishkek via Dushanbe.

Alpinism as a Spatial-Agent-System

For alpinism, relatively extensive logistics are needed. They include supply at the place of departure, access to the base camp (today mostly with modern aircraft) as well as the organisation of the expedition itself including the involvement of local guides. Mostly, equipment and food have to be organised in advance in the cities. The numbers of tourists for alpine pursuits are not high – particularly because the season lasts not more than two months in summer. As various services are demanded, a considerable number of jobs and locations are however at least partially involved in the system. Thus, alpinism is linked with relatively high expenditure per person and day. Some tour operators have specialised in this segment of tourism and cooperate with foreign partners so that the value added chain reaches from services in the alpinists' home countries to local services in Kyrgyzstan and Tajikistan. A significant factor for alpinism is the publication of reports in scientific and popular magazines, larger expeditions have media and TV presentation, and recently on the Internet, which is becoming increasingly important.[7]

Community-Based Tourism: From a Concept of Rural Development to an Organised Tourism System

In an analogy to projects in Nepal, the concept of community-based tourism (CBT) was initiated in Kyrgyzstan in 1999 as a means of delivering sustainable rural development. Ideally, CBT is a form of tourism where the local service providers are involved in tourism development right from the start and play a significant role in its management. Besides the factor of participation, CBT seeks to improve the socio-economic conditions of the community, aims at conserving and reviving historical and cultural resources and is concerned with the conservation of natural resources. Thus, the implementation of the concept contributes to the achievement of the targets of Agenda 21, which was revealed at the 1992 UNCED in Rio de Janeiro and of the Millennium Development Goals.

Promotion of Tourism as a Task of International Development Cooperation

Projects of international development and cooperation were the general framework of this tourism concept in Kyrgyzstan. Given the economic and political conditions in the

---

[7] Unfortunately, no statistics are available on either alpinism in Kyrgyzstan or the value added chains connected to it.

country, only minor public investments can be expected in the promotion of tourism so that either private investors have to be encouraged or low cost forms of tourism have to be developed which are profitable, but which still aim at sustainability.

Helvetas, a Swiss development agency operating in Kyrgyzstan since 1994, aims at increasing the capacity of the rural population to initiate and sustain its development towards improved living conditions. As part of its programme, Helvetas initiated the *Community Based Tourism Support Project* (CBTSP), which has developed into the most extensive and most important tourism project in Kyrgyzstan. It started in 1995 as the *Women's Promotion Project* (WPP) familiarising rural women in the Naryn and Ysyk--köl oblasts with the principles of market economies and motivating them to start or improve their own small business projects. The first CBT group was founded in 1999 in Kochkor and in the following years the project expanded to ten groups throughout the country so that in 2003 the *Kyrgyz Community Based Tourism Association* (KCBTA) was registered as a membership-based association of CBT groups in Kyrgyzstan. In 2004, *Shepherd's Life*, an NGO sourced out of a Helvetas development project in 1997, was integrated into KCBTA. *Shepherd's Life* has initiated a concept where tourists should have the opportunity to share the simple life of shepherds for a couple of days at the *jailoo* (high mountain pasture-land) and thus to get an impression of the daily life in a remote mountain society. *Shepherd's Life* shares the targets of CBT, including a sustainable management of natural resources, the conveyance of local knowledge and a better understanding of the traditional way of life and economic situation of the local population.

Currently, KCBTA consists of 17 CBT groups and Shepherd's Life. The CBTSP ended in 2005 so that all project responsibilities were taken over by KCBTA. So far, Helvetas continues to provide financial support to the KCBTA. It is hoped that KCBTA will be financially independent in 2009.

An important target of Helvetas and KCBTA is to give the local CBT groups comprehensive support in the area of marketing, training and organisational development. CBT groups are self-managed membership organisations and receive tourism- and business related training from CBTSP to develop into viable and well-managed local organisations. This is continued by KCBTA, which offers a wide range of workshops ranging from bookkeeping training, pro-poor tourism training to welcome-guest-courses for home stay owners. Local tourism information centres were set up in the CBT groups to provide the tourists with information on accommodation and further services and to organise the distribution of tourists to the CBT members.

In the course of time, the system of CBT became more professional. In order to better monitor the quality of CBT services, KCBTA has introduced an accreditation system and developed a gradation system for home stays. Three different standards in combination with graded prices offer incentives for the tourism providers to improve the standard of their services and provide tourists with information on the standard that can be expected. The linking of CBT groups with the help of KCBTA makes it possible for tourists to use CBT services during their travels nationwide.

In some places, spin-offs can be found. Their operators are usually still in contact with CBT and due to a lack of own marketing, they are dependent on positive knock-on effects of the local CBT group. An example for this is the *Jailoo Tourism Community* in Kochkor, which was initiated by a former CBT member. By now, the Kyrgyz CBT concept has developed into a role model for sustainable tourism development in Central Asia and has encouraged the development of similar organisations in the neighbouring countries.

As accompanying measures to the CBT project, Helvetas has initiated the Destination Marketing Organisation Project, which aims at improving market access for local producers and increasing the degree of awareness of Kyrgyzstan as a tourism destination. It is intended to position Kyrgyzstan as a brand on the international tourism market. The destination marketing focuses on two main target groups: recreational tourism is mainly promoted in the Russian and Kazakh tourism markets whereas marketing of adventure and cultural tourism is primarily concentrated in western countries. A positive effect ranging from a national to a local level can be expected.

KCBTA closely cooperates with tour operators in Kyrgyzstan; in 2006 framework agreements were signed with 16 tour operators. Among them is NoviNomad, a company for tourism development that was initiated by Helvetas and set up by a former CBTSP consultant. NoviNomad has adopted the goals of sustainable mountain tourism, arranges trips to various destinations in the country with accommodations in CBT homes and cooperates with local CBT groups by jointly organising festivals. About 150 additional people are involved in the work of NoviNomad during the tourism season.

Case Study: Kochkor and the Summer Pastures at the Son Köl as CBT Destinations

Kochkor in central Kyrgyzstan exemplarily displays the development of CBT in Kyrgyzstan and shows how it operates, its social implications and several of its economic effects.[8] Kochkor is a rural community at a crossroads where the main road from Bishkek to Naryn, the centre of the administrative region, crosses a track leading from the south shore of Ysyk-köl to the west in the direction of Chaek-Susamyr where it connects to the north-south track from Bishkek to Osh. Due to these transport connections, Kochkor initially became a market town and gradually developed into an administrative and economic centre (STADELBAUER 2006). With ca. 15,000 inhabitants and a shopping centre, Kochkor has an urban structure nowadays and is now considered to be one of the most important transfer hubs for tourists in the country – on the one hand because of its good connections, on the other hand because of its location between numerous attractive high mountain pastures which turn Kochkor into the most important base for *jailoo* tourism.[9]

---

[8] The following analysis is based on a study carried out as part of a master thesis (SCHNEIDER 2007) which was financially supported by the DAAD. A comparable study was undertaken in Arslanbob in the context of a research project in Erlangen and Berlin funded by the Volkswagen Foundation (KIRCHMAYER & SCHMIDT 2005).
[9] *Jailoo*-tourism refers to stays of tourists in yurts on the summer pastures. In addition to being able to stay in private yurts, which offers insight into the families' daily life, special yurt camps, consisting of several yurts, can house bigger parties.

The development of *jailoo* and CBT tourism in Kochkor only started after Kyrgyzstan became independent. It was initiated by the programme Shepherd's Life in 1997 and was followed by the establishment of Kyrgyzstan's first CBT group in 2000. Among others, the realisation of CBT objectives revived handicraft skills. This includes the erection of a yurt and the fabrication of felt carpets (*shyrdak*) as well as the production and use of fermented mare's milk (*kymys*). Kochkor also has the function of a role model in the sphere of handicrafts: in 1998 the NGO Altyn kol ("golden hands"), a women's handicraft cooperative producing traditional Kyrgyz handicrafts, such as felt rugs, was formed in Kochkor and has now more than 200 members. At the same time, the local CBT group developed further. At the beginning, the group only focused on the provision of accommodation and food. Later, the CBT group also offered the service of guides and translators as well as renting horses which led to the involvement of further agents in the network of CBT and positive social effects (Fig. 3). This working framework now includes large parts of the community and its surroundings.

A special advantage of the location of Kochkor is the access to Son Köl, a mountain lake situated in a shallow depression at an altitude of 3,000 m altitude, surrounded by traditional summer pastures. Although the Son Köl is part of a biological reserve, it is still used by nomads. They stay at the summer pastures with their herds from the middle/end of May until September, sometimes even until the beginning of October before they return to their villages in the valleys for winter. The peaceful quiet landscape, high biodiversity, quickly changing weather, the possibility for horseback tours and hiking and the remoteness of the lake make Son Köl very attractive for tourists. It is therefore the most popular destination for *jailoo*-tourism originating at Kochkor. On average, tourists stay for one night in Kochkor and for three nights at Son Köl. It is estimated that in 2001 about 1,200 tourists visited Son Köl with a slight increase of 5-10 percent in the following years (DUDASHVILI 2005: 54).

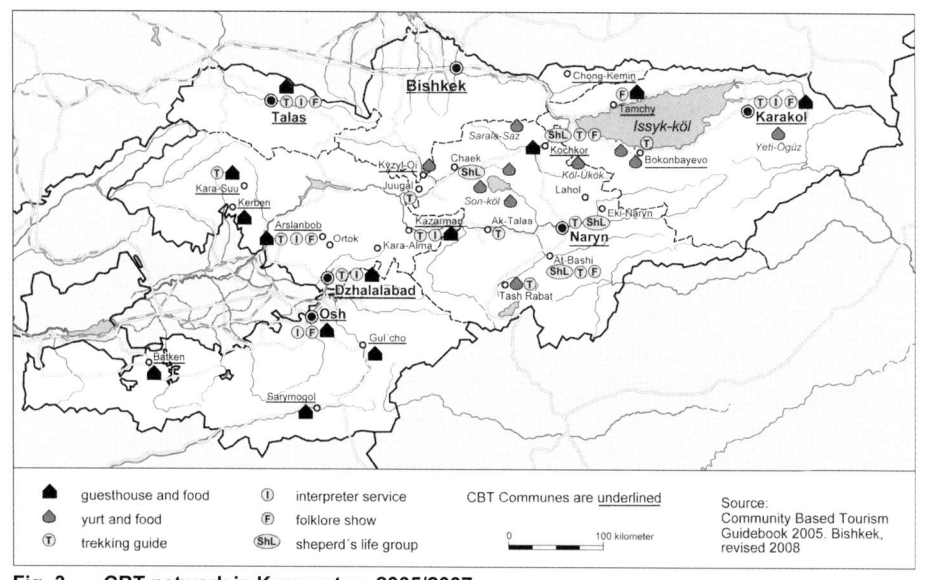

Fig. 3   CBT network in Kyrgyzstan, 2005/2007

The economic implication of CBT for the community can only be roughly estimated. In 2006, CBT Kochkor had an income of about US$ 45,000 which equals a daily per capita spending by the tourists of US$ 26.82 (field survey 2006). 15 percent of the money earned by the members goes to the local CBT organisation to cover running expenses and is also used for ecological and social projects in Kochkor. In terms of income generated and number of visitors, CBT Kochkor is the most successful CBT group in Kyrgyzstan. However, its positive effect on the local labour market should not be overestimated. CBT does create additional jobs, but these are mainly seasonal and concentrated in the three month long summer season. The indirect and non-monetary implications of CBT are more important and show its role as having a positive multiplier effect on the local economy and society. Such as seen for example, in follow-up investments of private households and the revival of old traditions and customs. The improvement of living standards only develops step by step but is perceived by its members. Social aspects are considered to be as important as economic effects. For women in particular, the involvement in a CBT group creates new opportunities and responsibilities where new ways of cooperation and social contacts become important.

CBT as a Network in Rural Regions

To sum up, the CBT concept can be considered as a successful development strategy in remote rural regions. So far, CBT has achieved its aim of increasing income in rural communities and developing a new tourism product for the international tourism market. An ecological code of conduct created for the CBT groups aims to raise ecological awareness. The aspect of intercultural contacts and exchanges between hosts and guests is also very important, as it helps to preserve traditional crafts and offers authentic insight into Kyrgyz life for the tourists. However, the CBT concept is no "silver bullet" for rural development and expectations should not be set too high. The monetary flow to the members is so far still modest. Part of the income goes to the CBT office and creates some permanent employment and local linkages. Furthermore, an increase of income for local service providers outside the CBT group can be identified. However, there is the danger that the social, cultural and ecological aspects of the CBT concept could be neglected in favour of the economic success of the group.

Thus so far, community based tourism has not been able to solve all existing problems and address all shortcomings in rural communities nor to trigger a prospering development of the national tourism industry. Yet, for the mostly socio-economically weak parts of the population involved in CBT, it creates a chance to alleviate poverty. Although CBT is only a relatively small segment of the tourism in Kyrgyzstan, it is an essential contribution to the realisation of the UN Millennium Development Goals. Particularly in remote rural regions, CBT can be understood as a form of pro-poor mountain tourism and is thus respected not only in Kyrgyzstan *(ICIMOD* 2007: 41 ff.)

Tab. 1 CBT at Kochkor: Results and problems

RESULTS ACHIEVED	PROBLEMS
Economic	
<ul><li>Creation of jobs (directly and indirectly), particularly for women by offering accommodation and services such as guides, interpreters, transportation</li><li>Generation of income (though primarily for CBT members)</li><li>use of income for *community development projects* (percentage of income unknown)</li><li>role as a multiplier: - diversification, growth of local economy - use of local resources, traditional knowledge (production of felt handicrafts, nomadic life)</li><li>Investment - of members themselves (also possible because of loans from CBT) - of CBT into projects</li><li>Quality management by accreditation and certification of CBT services</li><li>cooperation with stakeholders of the private and public sectors, umbrella-organisations for better marketing</li><li>target group aligned marketing (own website) with further development to e-commerce</li></ul>	<ul><li>marginal additional value created (relatively small number of direct employment, little fulltime employment)</li><li>granting of loans only for members: exclusion of poorer parts of the population as the socio-economic situation hinders their access to the CBT</li><li>no continuity in maintenance of internet presence to foster market shares</li></ul>
Social	
<ul><li>improvement of living standards of the local community</li><li>fair division of roles between women/men and old/young</li><li>involvement of disadvantaged parts of the population (particularly of women)</li><li>social projects: also non-CBT members profit from tourism</li><li>training and workshops for tourism services</li></ul>	<ul><li>external and internal conflicts: - jealousy and distrust - rivalry</li><li>formation of a new elite (not open to everyone, the poorest are excluded)</li><li>danger of fusing municipal and tourism services</li></ul>
Environmental	
<ul><li>raising of awareness of the need for conservation in the local community (by tourists, workshop on environmental issues)</li><li>Management of waste disposal (clean up actions at Son köl and other *jailoos*)</li><li>sustainable relationship between tourism and nature (ecological codes of conduct available)</li></ul>	<ul><li>danger of overusing natural resources and degradation of pasture land</li><li>no monitoring if the codes of conduct are implemented, insufficient information for tourists</li></ul>
Cultural	
<ul><li>cultural exchange</li><li>respect for different cultures</li><li>promotion of local culture, revival of traditional ways of life and customs (nomadism, production of felt handicrafts, etc.)</li></ul>	<ul><li>acculturation (partial adoption of western values)</li><li>danger of lasting misunderstandings</li></ul>
Political	
<ul><li>participation of local people (individual responsibility of members due to participation in regular meetings, projects, committees)</li><li>strengthening of CBT group to the outside (CBT group is acknowledged as an important regional NGO by politicians)</li></ul>	<ul><li>participation of whole community due to high number of inhabitants not possible</li></ul>

source: field survey, Andrea Schneider 2006

CBT as an Element within a Mountain Socio-Economic System

Which systemic interdependencies/interfaces does the CBT strategy have? A look at the achieved aims and existing problems shows elements of the concept (tab. 1, Fig. 3):

The success of the whole system depends on two feedback loops:

- The system has to be designed as a constant and sustainable concept, including stable structures within the local groups and the umbrella organisation as well as continuous marketing at a high level that is appropriate for its target groups.[10]
- There has to be continuous self-monitoring (*ICIMOD* 2007, 1: 118 ff.). So far, it is difficult to say whether Kyrgyzstan has already developed in this direction.

On one hand, the latent perception of Central Asia as a region troubled by conflicts remains a negative factor in the international marketing of Kyrgyzstan as a tourist destination. On the other hand, the development of tourism in this region is considered to be a key contributor to poverty reduction – regardless of the problems, which can result from social disparities between guests and locals.[11] It is expected that the implementation of sustainable tourism focused on regional development with close linkages to agriculture, small trade, and in some regions to forestry and the management of biological reserves, will be a successful instrument of poverty reduction in remote rural areas.

The Spatial system of Agents in Tourism in the Ysyk-köl Region

General Conditions

The present spatial system of agents in tourism in the Ysyk-köl region (cf. most recently ALLEN 2006) is based on various location factors of the lake shore and its surroundings: It is the second largest mountain lake in the world and due to its continental location, it is a convenient alternative to seaside holidays. There is no comparable location in Central Asia where lake-side recreation can be combined with mountain tourism in such a way. The settlements around Lake Ysyk-köl can be reached within a few hours from the Kyrgyz capital, Bishkek, and in the summer there are direct flights from Almaty, Kazakhstan, to Tamchi on the north shore of Lake Ysyk-köl. Border controls at the Kyrgyz Kazakh border are sufficiently relaxed to allow relatively easy access to visitors from the conurbation of Almaty. This development of tourism goes back to a time when Almaty was still the capital of Kazakhstan. The relocation of the capital to Astana does not seem to have affected the presence of Kazakh tourists at Ysyk-köl.

---

[10] This also includes continuous accessibility on the internet which is not always possible in the case of Kyrgyzstan. As an example, the website of KCBTA was not always available at the beginning of 2008 although they had registered an increase of visitors on their website from 15,420 in 2005 to 50,590 in 2006 (*KCBTA* 2006: 23).

[11] Negative consequences of tourism in developing countries include a lack of attention to social conditions, disregard of customs and traditions, profanation of holy places, alienation of the local population from their traditional way of living, additional work particularly for women who mostly have to host the guests.

In Soviet times, there were already plans to develop the shore of Lake Ysyk-köl for mass tourism (Fig. 4). Recent proposals, although not completely aligned with the plans made at the end of the 1960s, do not envisage development occurring in a completely different manner from these earlier strategies. The most important difference is the new framework of decision-making, which was created by privatisation in the tourism infrastructure. Further incentives for tourism at Ysyk-köl were created by the attempt to initiate a biosphere reserve. In this context, the GTZ has tried to introduce a concept of sustainable tourism (*GTZ* 2002). However, a first critical analysis has stressed the difficulties existing in the implementation of the concept (HÜNNINGHAUS 2001).

The current tourism strategy emphasising water-based tourism is problematic as the tourism season for bathing is, depending on the weather, limited to only 60 to 70 days per year. Therefore, strategies for an extension of the season are recommended, such as the construction of bathing facilities which can be used year-round and incorporate not only swimming pools, but also thermal baths, special child-friendly bathing pools, etc and other facilities which can be used all year round (*IMF* 2007: 38). So far, this has failed due to high costs and the lack of investors.

Structures

With 890 establishments providing tourist accommodation, the Ysyk-köl oblast has the highest number of providers of all Kyrgyz oblasts. In the Ysyk-köl region, about 5,000 people are employed in tourism – this is about 50 percent of all people working in tourism in Kyrgyzstan. This intensity highlights the exceptional position of the area along the lake shore (*Uslugi…*, 2005: 14).

At Ysyk-köl, private initiatives play a vital role. Although a regional classification is not possible with the available data, a high number of the total of 12,200 registered beds is expected to be located on the north shore of the lake. Out of the 709,100 visitors, which were registered in 2004 in Kyrgyzstan, 235,800 (33.3 %) stayed in private accommodations. In the Ysyk-köl oblast, the number of visitors staying in private accommodations was even higher with 35.8 percent (207,400 out of 579,400). Visitors to Ysyk-köl come not only from Kyrgyzstan itself, but also to a great extent from Kazakhstan, which is economically better off. Often, Kazakhs visit Ysyk-köl for a long weekend although the two routes from Almaty via Bishkek and Kegen are 497 km and respectively 460 km long. A more direct connection, which will be about 200 km shorter, has been under construction for a couple of years.

The centre of recreational tourism at Ysyk-köl is Cholpon-Ata on the north shore. It has already developed into a tourism centre but is still only in the early stages of developing a professional infrastructure. Nonetheless, it has further potential for development: a tourist information office is being set up in the Ecocenter. However, a map of Cholpon Ata is still not available and tourist information materials about the town, accommodation facilities, etc. are sparse. The Ecocenter was built in 2004 as part of a development project setting up the biosphere reserve at Ysyk-köl. It is, however, unclear whether the plans to use the Ecocenter as a multifunctional facility can be fully realised in future. Furthermore, it has to be considered that Kazakhs from Almaty make up the highest percentage of tourists at Ysyk-köl – a target group that has so far showed only little interest in ecological regional development. The renting of bedrooms, restaurants, cafés, souvenir shops, currency exchange offices and various other facilities have an important effect on employment, although tourism is restricted to the summer season. Due to the weather and the summer holidays, August is the most popular time for tourism at Ysyk-köl. This is also reflected in pricing of accommodation and transport. In 2006, for example, a trip with a shared taxi from Cholpon-Ata to Bishkek cost 250 Som in August and only 150 Som in September.

The spatial-agent system in the Ysyk-köl region is not primarily focused on mountains – unless one defines the Ysyk-köl and its direct surroundings at an altitude of about 1,900 m as high mountains. Instead, tourism at Ysyk-köl forms a system focused on recreation involving many stakeholders and participants. The surrounding mountains and the possibility for trekking tours are only a secondary aspect of the recreational system – not its focus. Offers of excursions to other attractive sites spatially embed further sites at the shore and the neighbourhood into the tourism system. Yet so far, the potential for tourism development at Ysyk-köl is used only insufficiently because the opportunities and possibilities provided by tourism that is focused on regional development are still used only marginally.

Tourism in Kyrgyzstan in the Context of Development of Mountain Regions – Some Problems

The development of tourism is embedded in the general development of countries and regions. As an economic system, it is part of a value added chain, including upstream and downstream linkages and affects many more people than those directly involved in

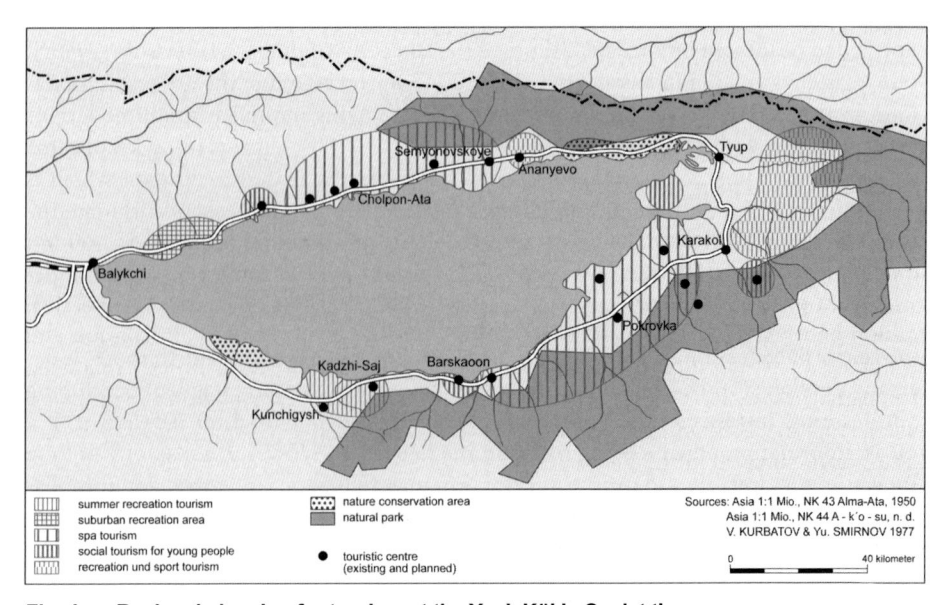

Fig. 4   Regional planning for tourism at the Ysyk Köl in Soviet times

tourism by its multiplier effects. At the same time, tourism development is integrated in a social system which goes beyond economic linkages, is not always free of conflicts (tradition vs. modernisation) and can imply affective categories such as prestige, appreciation or enviousness, particularly at a local level. Furthermore, tourism development is embedded in a spatial system that not only includes the immediate locations of tourism activities, but also wider areas through its linkages to subcontractors, as well as a range of economic operations in the infrastructure and by the people involved.

Despite its success at a local level in some areas of Kyrgyzstan, tourism so far does not have the same impact and economic importance as it has in other mountain countries. What are the reasons for this? Economically, tourism in Kyrgyzstan has only recently started to be part of the value added chain. First attempts to achieve this are the linkages between newly founded travel agencies, service providers of local tourism organisations and transport companies as well as establishing cooperation with institutions concerned with developmental cooperation and research. By integration in long-term public development strategies and international programmes, tourism concepts also gain a political component. The EU Central Asia Strategy, which was passed in 2007, however, does not list tourism as a primary development goal.

Tourism development in mountain regions does not proceed without conflicts as tourism might be in competition with other uses, such as grazing.[12] However, previous studies on Kyrgyz pasture land (BLANK 2007) have not mentioned such problems so far. In the case of Arslanbob / Kyzyl Unkür, for example, there is only limited interface between pasture farming and tourism.

A natural handicap for the further development of tourism in Kyrgyzstan is the restriction of tourism to a short season in summer. So far, strategies to expand the tourism season are missing. The preparation of Kyrgyzstan's tourism destinations for the international market is still in its infancy. Kyrgyzstan regularly participates in international tourism fairs, such as the ITB in Berlin, and is increasingly present on the Internet, but a broad marketing strategy is still lacking. The size of CBT groups, with at most about 30 members, depends on demand; the groups only expand in areas with an increasing demand. Mass tourism is not part of the CBT concept.

In Kyrgyzstan, improvements in infrastructure and logistics are required for further development. The IMF considers tourism development to be one among several development strategies for the country and expects investment needs of about US$ 110 million between 2007 and 2010 (*IMF* 2007: 41). Furthermore, de-politicisation and simultaneous stabilisation of the political situation is necessary. The unrests following the change in the presidency from Askar Akayev to Kurmanbek Bakiyev had the consequence that in 2005, only about 320,000 foreigners visited the country, after there had been nearly 400,000 in the previous year.

The creation of real participation in tourism is still difficult due to inherited structures of decision making. Particularly, the role of women in tourism development might be

---

[12] Described for the Pyrenees by MARIN-YASELL & MARTINEZ (2003): better working conditions since the tourism boom in the 1960s, higher income, decrease of cattle, decrease in pasture due to building development on former grazing land.

of special interest in further analysis: on the one hand there is a lack of (paid) jobs for women in central Kyrgyzstan, on the other hand tourism concepts which aim at the participation of individual households often lead to additional work for the women who already have to keep the household and work on their small farms.

The strategy of Kyrgyz tourism policy still has to be implemented. A part of the strategy is the government programme "Development of the tourism sector of the Kyrgyz Republic until 2010" which was passed in 2000. Its five key aims are the protection of natural and cultural heritage resources, the creation of jobs, increases in income, the stimulation of other economic activities and the increased foreign investment (THOMPSON & FORSTER 2003; ALLEN 2006: 103 ff.). As a success for the Kyrgyz government, it has to be mentioned that the year 2002 was designated International Year of the Mountains by the UN, following a proposal of the president, with the Bishkek Global Mountain Summit as its final conference in October 2002. This event also emphasised the importance of tourism for remote rural regions. So far, it is not clear which parts of the old tourism strategy will be continued by the new government under Bakiyev.

The three types of tourism presented here all operate in different spheres. They all have positive economic and social effects but tend to operate in isolation rather than co-exist in an integrated system. Close spatial and structural linkages are still lacking – they could facilitate combinations of various tourism activities and the specific demands and requirements of target groups. Tourism in Kyrgyzstan remains far away from developing its own brand with a diversified range of tourist services and attractions.

References

ALLEN, Joseph Boots (2006): What About the Locals?: The Impact of State Tourism Policy and Transnational Participation on Two Central Asian Mountain Communities. Diss. Phil. Austin/Texas. <https://www.lib.utexas.edu/etd/d/2006/allenj39461/allenj39461.pdf>

ASYKULOV, Tolkunbek; Matthias SCHMIDT (2005): Naturschutzkonzepte im Transformations-prozess: Das Biosphärenreservat Ysyk-köl in Kirgistan. Natur und Landschaft 80 (8): 370-377.

BLANK, M. (2007): Rückkehr zur subsistenzorientierten Viehhaltung als Existenzsicherungsstrategie. Hochweidewirtschaft in Südkirgistan. Berlin (= Occasional Papers Geographie; 34).

DUDASHVILI, Sergei (2005): Turizm v Kyrgyzstane, Bishkek.

GTZ (2002): Investment-Handbuch für eine nachhaltige Tourismus-Entwicklung am Issyk-Kul. Eschborn u. Bischkek.

HAHMANN, Thomas (2007): Multi-Scale Planning and Design of a High-Mountain Long-Distance Trail. The Example of the "Kyrgyzstan Trail" in the Tien Shan Mountains. Unveröff. Diplomarbeit, Technische Universität Dresden, Studiengang Kartographie.

HÜNNINGHAUS, Anke (2001): Management von Biosphärenreservaten in Transformationsländern, dargestellt am Beispiel des Biosphärenreservats Issyk-köl in Kyrgyzstan. Diss. geowiss. Bochum.

ICIMOD [ed.] (2007): Facilitation Sustainable Mountain Tourism. Vol. 1: Resource Book; vol. 2: Toolkit. Kathmandu.

IMF (International Monetary Fond) (2007): Kyrgyz Republic: Poverty Reduction Strategy Paper – Couuntry Development Strategy (2007 – 2010). Washington (IMF Country Report No. 07/193).

KCBTA [ed.] (2006): Kyrgyz Community Based Tourism Association „Hospitality Kyrgyzstan". Yearly Report 2006. Bishkek.

Kirchmayer, Carola (2005): Tourismus im Transformationsprozess. Entwicklung und Wirkungsfelder in Arslanbob (kirgistan). Erlangen, unpubl. master thesis.

Kirchmayer, Carola; Matthias Schmidt (2005): Transformation des Tourismus in Kirgistan: Zwischen staatlich gelenkter rekreacija und neuem backpacking. TourismusJournal 8 (3): 399-417.

Kurbaton, V.; Yu. Smirnov (1977): Novyi krupnyi kurortnyi rayon na ozere Issyk-kul'. Architektura SSSR 1977/4: 26–33.

Marin-Yasell, M. L.; T. L. Martinez (2003): Competing for Meadows. A case study on tourism and Livestock Farming in the Spanish Pyrenees. Mountain Research and Development 23 (2): 169–175.

Natsional'nyi statisticheskii komitet Kyrgyzskoi Respubliki (2007): Kyrgyzstan v tsifrakh 2007. Bishkek.

Preobrazhenskii, V.S.; V. M. Krivosheiev (red.) (1980): Geografiia rekreatsionnykh sistem SSSR, Moskva.

Schneider, Andrea (2007): Tourismus als Entwicklungsstrategie im ländlichen Raum Kirgisistans, Wissenschaftliche Arbeit für die Prüfung für das Lehramt an Gymnasien, Freiburg i. Br., unpubl.

Schneider, Andrea; Jörg Stadelbauer (2007): Auf der Hochweide in Kirgisistan. Lokaler Tourismus und Regionalentwicklung. Osteuropa 57 (8/9): 559–566.

Stadelbauer, Jörg (2006): Struktur und Entwicklung des Städtesystems in einem Gebirgsstaat – das Beispiel Kyrgyzstan. In: Gans, P., A. Priebs, R. Wehrhahn [eds.]: Kulturgraphie der Stadt. Kiel 2006 (= Kieler Geographische Schriften; 111): 587–604.

Thompson, Karen; Nicola Forster (2003): Ecotourism Development and Government Policy in Kyrgyzstan. In: Fennell, D. A.; R. K. Dawling (eds.): Ecotourism Policy and Planning: 1691–1789.

Uhlemann, Kathrin [coord.] (2003): Biosphärenreservat Issyk-kul. Inventar der kulturhistorischen Stätten, Eschborn, Biškek.

Political Ecology in High Mountains: the Web of Actors, Interests and Institutions in Kyrgyzstan's Mountains

Matthias Schmidt[1]

Introduction

The varied relief of mountain ranges in its horizontal and vertical dimensions, its wide range of different micro-climatic conditions due to elevation and exposure, its manifold geologic formations and geomorphologic aspects all imply a high diversity of ecological conditions. Mountains contain various minerals and are covered with multifarious floristic structures and different types of forests or grasslands. But such natural entities are appreciated as resources only because of human demand. Humans endow value to minerals, biotic elements or sheer location. They articulate concerns about specific natural products or mountain territories for various aims.

Consequently the heterogeneous nature of mountain ranges can make them into hotspots of diverse concerns expressed by manifold actors. The scope of interests in mountain areas is wide: it includes not only economic or political issues, but also social, cultural, environmental or recreational concerns, and all of them are connected with specific human actors. Admittedly, not only the local mountain population, the 'place based actors', are involved in using mountain territories or natural resources in mountains; also people not living or interacting directly on the spot, the 'non-place based actors' (BLAIKIE 1985), decide about the exclusion of possible stakeholders or the form of resource management, for example, or demand specific resources or functions of the mountains.

Since natural resources and mountain territories are limited, access, usage and control of them by specific actors are the result of political negotiation processes, in which power positions and power relations are manifested. Environment is thus to be seen as battlefield of diverse concerns, on which various actors fight for power, usufruct rights and influence (BLAIKIE 1995). Likewise mountains are arenas in which different actors located at, or related to, different scales struggle to enforce their aims.

The goal of the present analysis is to understand how particular land use and land management regimes evolved through the intersection of ecological, political, economic and social structures and developments. With the example of environmental change in a small mountain area in Kyrgyzstan, the degradation of walnut-fruit forests in the Western Tian Shan, I intend to show the varied web of involved actors, their concerns, and the regulating institutions against the background of a historical analysis.

Resource Utilisation and Human-Environmental Relations in High Mountain Research

Until the 20th century mountain regions were mostly characterised as distinct and peripheral islands. Research on mountains concentrated on ecology and physical features.

[1] Freie Universität Berlin, Institut für Geographische Wissenschaften; D-12249 Berlin, mschmidt@geog.fu-berlin.de

The dominating geoecological approach of TROLL (1972) sought a greater understanding of the complex interactions of climate, soils, fauna and flora. In particular German geographers engaged in comparative high mountain research (TROLL 1975; UHLIG & HAFFNER 1984; RATHJENS 1988) and developed their own classification systems in the 1970s and 1980s. They pointed out specific characteristics of mountain areas such as vertical stratification of utilisation stages (UHLIG 1976; SOFFER 1982; SCHWEIZER 1984), high dependence on natural features and resources, low development stage, traditionalism, high mobility, and frequency of common property regimes (JENTSCH 1984), or specific settlement characteristics (GRÖTZBACH 1982). The vertical stratification of socio-economic zones was seen as definitive in the mountain context, analogous with the altitudinal zones of the physical geographers. Human geographers interpreted mountains as frontier areas to which humans have to adapt, and as additional resource areas offering specific products and functions to the more progressive forelands (RATHJENS 1982). The sometimes simplistic explanations of human adaptation to mountain environments show an environmentally deterministic thinking and were based mainly on a systems approach rather than on detailed studies of human interactions with their environment.

Within the frame of the 'Man and Biosphere' programme, scientific activities in mountain areas were intensified to investigate environmental, economic and cultural processes in the European Alps (UNESCO 1974). The project relies on a systems approach, which assumes that human communities are in balance with their environment and places little emphasis on people and their role as active agents in the environment (SMETHURST 2000: 38). Aspects of human impacts on mountains moved to the centre of attention and culminated in the discussion of the Himalayan Dilemma (ECKHOLM 1975). This theory is not only characterised by ecological ideas and elements of Neo-Malthusian collapse resulting from population increase and limited resources but also simplistic because the mountain farmers are treated as homogenous by ignoring historical, ethnological and socio-cultural distinctions which are relevant for conflict negotiations and power relations, marginalisation and impoverishment (KREUTZMANN 1993: 13). Although mountain farmers are seen as active agents, neither their concerns and constraints nor the influence of external agents are taken into consideration. IVES & MESSERLI (1984) see the "stability and instability of mountain areas" to be in danger from human influence. Population growth, increasing use of marginal land, overgrazing and deforestation were seen as the main harmful activities in developing countries, whereas tourism, recreation, forestry, and the building of roads and dams harm mountains in developed countries (GERRARD 1991: 75). Thus, most mountain studies of this period (cf. PRICE 1981) rely on methods drawn from the ecological sciences and show the handwriting of the quantitative revolution, with the growth of use of scientific data, modelling and quantification of parameters. Mountain areas are widely seen as closed containers with an in- and an outbox, though this perception of secluded mountain areas was already challenged in 1986 by ALLAN who criticises the altitudinal zonation model and proposes instead a model incorporating accessibility features taking into account the changing traffic infrastructure. HEWITT (1988: 22) even calls "into question the whole idea of mountain regions as a meaningfully separate area of investigation" because changes in mountains are dependent upon developments, initiatives and penetration from outside the mountains, which in turn influence the responses of mountain people. The importance of so-called 'highland-lowland interactions' and the integration of mountain regions into the global economy were thus stressed by several authors (Hewitt 1992, Kreutzmann 1993; Ehlers & Kreutzmann 2000).

Mountain regions received specific attention in chapter 13 of the Agenda 21 of the 1992 UN Conference on Environment and Development in Rio de Janeiro. Thus, mountains offer globally important resources such as water, energy, biodiversity, raw materials, wood, agricultural products, or landscape for recreation and tourism. Their importance, especially for the forelands, and the threats towards mountains were stressed (Stone 1992). Sustainable development became the new buzzword of almost all publications on mountain developments. But all areas of concern remain notably sterile and technocratic. This holds true when Winiger (1992) calls for modelling the whole mountain system and the change of human-environment relations, or if we consider the more recent publication "Global change and mountain regions" by Huber et al. (2005) in which only 125 of 650 pages are devoted to human dimensions (of global change in mountains), and political and institutional aspects are the focus of only two of sixty-one chapters. In such publications, humans are seen as calculable factors or elements of a system that can be measured, classified and modelled.

The decade following the Earth Summit can be interpreted as the decade of labelling mountains with terms such as limited accessibility, verticality, fragility and marginality, diversity and human adaptation mechanisms (Messerli & Ives 1997; Jodha 1997 315; Sarmiento 2000). Although the dynamics between humans and their environment were investigated and better understood in many ways, the culture of mountain peoples, political factors and processes within mountain areas and between mountains and surroundings are underrepresented in the mountain literature. For future research and monitoring of mountain areas Price (1999) and Kreutzmann (2001) demand the consideration of economic and social key indicators for mountain regions.

To sum up, there is still a need for mountain studies that include political, economic, cultural, and social dimensions of actors and their environments. Such studies need to include regional, national and international patterns of influence, the importance of mountain regions within national states and within the global economy, population dynamics and mobility, pressure on natural resources and land utilisation (Dikau et al. 2002 84). The consideration of human agents and their political positions, historical evolutions, external interventions and institutional frameworks is a priority issue.

Political Ecology in High Mountains

Political transformations, economic transitions, and penetration of international markets, foreign technologies and values within the recent globalisation processes as well as developments in the social sciences in general and in human geography in particular call for a different perception of mountain areas. The human-environment nexus remains central, while economic development, politics and power relations, livelihood issues and vulnerability (Watts & Bohle 1993; DFID 2000), social and ecological resilience (Adger 2000; Berkes et al. 2003; Berkes 2007) have emerged as particularly salient. Only few theoretical and methodological approaches are "able to understand these factors as integrated and variable over time, space, scale and specific context" (NIGHTINGALE 2003 525). The research agenda of political ecology offers an adequate tool for dealing with such questions. Political ecology is a theoretical framework for analysing shifting, dialectical relationships between social and power relations, autochthonous practices, and ecological processes to allow an interdisciplinary, complex assessment of social and environmental change (BLAIKIE 1985; BLAIKIE & BROOKFIELD 1987; BRYANT & BAILEY 1997; KRINGS & MÜLLER 2001; FORSYTH 2003; ZIMMERER & BASSETT 2003; ROBBINS 2004; NEUMANN 2005).

Political ecology, defined as an interdisciplinary field that combines "the concerns of ecology and a broadly defined political economy" (BLAIKIE & BROOKFIELD 1987: 21), offers an understanding of the reasons for social and environmental change in mountains. In this notion, environmental change is not only a matter of ecology but is linked with transformations of the political economy and thus connects the local arena of land use with decision makers and processes at regional, national or even global levels. The main premise of political ecologists is that ecological problems are at their core social and political issues, not technical or managerial problems, and therefore demand a theoretical foundation to analyse the complex social, economic, and political relations in which environmental change is embedded (NEUMANN 2005: 5).

The multi-scalar contexts of political ecology studies are important because human actors, their concerns in specific resources as well as management decisions are incorporated into various scales. Land users or land managers, the 'place-based-actors', are linked by 'chains of causality' with 'non-place-based-actors' (BLAIKIE 1985; BLAIKIE & BROOKFIELD 1987) in the wider society who affect them in any way. Property rights, control of, and access to resources were defined, negotiated, and contested at multiple scales. Exploitation of resources by impoverished land owners, for instance, is inextricably linked to political-economic processes operating at superior levels.

To analyse recent patterns of involved actors and concerns in mountain areas, historical analyses of the evolution of institutions, management practices and utilisation strategies are prerequisite. The study of environmental history and environmental discourses provides an understanding of the perceptions of local inhabitants and institutions (ZIMMERER 1993: 312). More attention to the dynamic nature of ecosystems and the historical contexts of social, economic and political regime change is necessary. PEET & WATTS (1996), BRYANT (1998) and many others have tried to theorise the linkages between capitalist development and land management. They figured out how capitalist development and productive relations influence land management regimes at different scales. Only in very rare cases have the relations between the socialist modernisation processes and environmental change been studied (SCHMIDT 2005a, b; HARTWIG 2007).

Social relations, economic practices and access to resources are governed by institutions.[2] Property rights, which should be defined as a bundle of rights, regulate access to and control of resources and land use, and thus the derivation of income from them. They include the right to possess, use, manage, alienate, transfer, and gain income

---

[2] According to NORTH (1990) institutions include rules, norms and traditions, as well as the organisations who establish, implement and control such regulations.

from property (SCHLAGER & OSTROM 1992). Discourses about ecology, land use and development are constituted, contested and reproduced through various institutions at different levels such as forest services, government planning, development agencies or environmental organisations. Rules, traditions, norms and property rights as well as organisations that enforce such regulations thus have significant implications for the resulting land management practices.

The general transition in most social sciences from positivism to post-structuralism as well as the preference for qualitative analyses needs to be reflected in mountain research, too. The uniqueness of local places, environmental perceptions and response must be taken into account. While environmental orthodoxies such as the Himalayan Dilemma theory reflect an essentially positivist and inferential approach to knowledge creation and explanation, post-structuralist approaches are more interpretative and stress the "individuality of environmental perception, and the injustice or inadequacy of uniform meta-statements which refer to everyone" (FORSYTH 1998: 110). Critical realism as a theoretical approach seems to be adequate for dealing with human-environmental relations in mountain areas because it states that environmental processes have an 'external reality' to human experience (BHASKAR 1986; SAYER 1992). This means that ecological processes must be seen as real and external to human experience, but that all knowledge claims about environmental processes are socially constructed. Discourses and contexts within which knowledge about ecology and land use is generated, how "nature" or "mountain systems" are constructed, or how social and economic practices are produced, in particular historical, cultural and ecological contexts must be more closely questioned (ESCOBAR 1996).

Political ecology studies can demonstrate the importance of political, social and economic relations, the evolution of institutions and present institutional arrangements, discourses about land use or environmental protection in analyses of environmental change. The example of Kyrgyzstan's walnut-fruit forests will serve to show the multifarious net of concerns, actors and transformed institutions against the background of significant political transformations.

Kyrgyzstan's Walnut-Fruit Forests: Spatial and Historical Setting

The study area is part of the Tian Shan Mountains and politically located within the Republic of Kyrgyzstan, which became independent in 1991 after the dissolution of the Soviet Union (fig. 1). The mountainous area is characterised by different ecological formations, of which the walnut-fruit forests, located at altitudes between 1000 and 2000 m at the south-facing slopes of the Fergana Range, are the most prominent. Major tree species of these forests, which cover an area of around 25,600 ha (MUSURALIEV 1998: 5), are walnut (*Juglans regia*), maple (*Acer turkestanicum*) and various fruit-bearing species in their wild form, such as apple (*Malus sieversii*), pear (*Pyrus korshinskyi*), plum (*Prunus sogdiana*), barberry (*Berberis oblonga*), rosehip (*Rosa kokanica*) and sea buckthorn (*Hippophae rhamnoides*) (cf. GOTTSCHLING et al. 2005). Steppe vegetation, pistachio groves and arable lands dominate in the areas below 1000 m, while shrubs, alpine mats and grasslands are predominant above 2000 m. The area is situated in the immediate vicinity of the densely inhabited Fergana Valley, one of the main economic

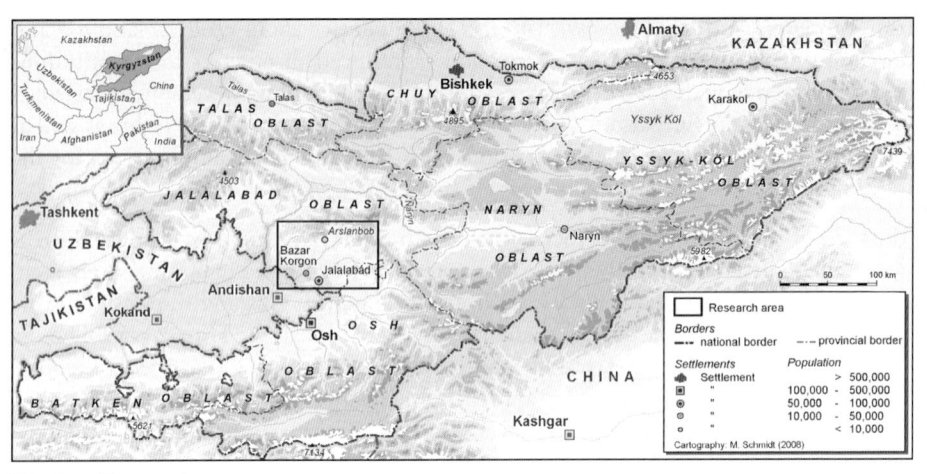

Fig. 1   Kyrgyzstan

areas of Central Asia. Because the ecological formations offer valuable resources such as timber, firewood, fruits, herbs and grass, these natural resources have come into the focus of human concerns at an early stage.

Various sources (e. g., LISNEVSKI 1884; KORZHINSKII 1896) indicate that utilisation of land and forest resources prior to the annexation of the area by the Russian Empire in 1876 was limited to the autochthonous population: nomads and sedentary populations who lived in a couple of small villages, cultivated fields, used the forests and grasslands as grazing grounds, collected fruits and produced charcoal, which was sold on the markets in the Fergana Valley. From 1889 to 1897 an expedition led by Russian explorers inventoried all potential natural resources of the wider Fergana region (NAVROCKII 1900). Initially, the walnut-fruit forests were highly esteemed not owing to their valuable products such as timber, nuts or fruits but because of their ecological function, especially their positive impact on the hydrology of the region, which was seen as essential for the long-term functioning of the irrigation systems in the Fergana Valley (RAUNER 1901). Consequently the Russian administration of the Governor-Generalship of Turkestan prohibited several forest usages such as felling, charcoal production or extension of arable land, and conceded to the local populations only the right to use grazing grounds according to their traditions (Svod zakonov Rossiiskoi Imperii 1892). However, the newly founded forest service noticed the occurrence and high value of walnut burls and assured itself economic profit: at the end of the 19th century already, a considerable number of walnut burls were being traded and exported to Marseilles, France, where they were used for furniture production (Direktor Lesnogo Departamenta 1902).

Apart from the above-mentioned ecological and economic interest in the forests the Russian administration tried to gain political control of the territory and created administrative units. First they established forest farms and thus subdivided the territory according to its utilisation potentials; then they formulated and implemented related property rights. All land and forests were declared to be the property of the Russian Tsar; only houses and their surrounding gardens were declared private property (Svod zakonov Rossiiskoi Imperii 1892). Since the autochthonous population could not purchase any other land – in contrast to the immigrated Russians – and because they were forbidden to cut wood or to transfer land into arable fields they were in fact ousted from their traditional way of life. The ban on charcoal production meant the loss of additional income, while the limitation of pasturage rights resulted in a decline of winter fodder for their livestock. Capital yields were only realised by the Russian government by selling walnut burls. Another important aspect with regard to environmental change was the increased demand for timber for the construction of railway lines and for Russian settlers who introduced the wooden house to the area where adobe houses were traditionally prevalent. The administration enforced the Russian legal system without consulting the autochthonous population; furthermore, Russians and locals were treated differently by the law. Indeed, the ecological significance of the forests was realised at a very early stage and has remained a kind of environmental meta-narrative up to the present day.

After establishing their power in Central Asia the Bolsheviks continued the forest policy of Tsarist Russia for over a decade before the collectivisation processes at the beginning of the 1930s marked a deep cut in the institutional frame: Kyrgyz nomads were forcibly settled and expropriated of their livestock, while large agricultural state (*sovkhozes*) and collective farms (*kolkhozes*) were established. All land was declared public property (Gosudarstvennyi oblastnyi archiv Jalalabad, f. 126, op. 1, d. 362). During the 1930s and 1940s many institutional changes were implemented (cf. MUSURALIEV 1998), but the function of the forests as resource areas especially for timber, nuts and fruits became more prominent, in particular when the forests were subordinated to the "Vitamin Industry Union" (*Sojuz Vitaminprom*) (DISTANOVA 1974: 13).

The forced collectivisation, repression and prosecution of so-called *kulaks* – in theory owners of large estates but in practice farmers with just a couple of cows and sheep more than the average, or persons who resisted collectivisation – have led to fear and alienation between population and governmental institutions. Although the forests should have been protected for ecological reasons according to the forestry law of 1921 (Gosudarstvennyi oblastnyi archiv Jalalabad, f. 806, op. 1, d. 32 i 4), the frequent institutional transformations prevented long-term forest management strategies, while the collectivisation process and the difficult time during World War II have led to heedless exploitation of land and forest resources.

In 1945 the Council of People's Commissars declared the walnut-fruit forests to be State Fruit-Forest Reserves with specific regulations for protection and utilisation (Gosudarstvennyi oblastnyi archiv Jalalabad, f. 76, op. 1, d. 18, l. 14; DISTANOVA 1974: 9). A few years later the central authority of the USSR subordinated the walnut-fruit forests to the ministry of forestry and decided to transform the *sovkhozes* concerned into governmental forest farms, the so-called *leskhozes*, which were given the responsibility of carrying out all forestry measures on the local level. The majority of the local population found employment in these forest farms from which they could sustain their livelihood. According to the forest's status as State Fruit-Forest Reserves the general aim of forest management was twofold: first, forest protection by control of usages and realisation of forestry measures, and second, forest utilisation including the extraction of timber, firewood, nuts, apples, plums and herbs which were sold to other state enterprises. Local households obtained the right to cut grass on specific plots in the forest area to gain winter fodder for their livestock. In the 1960s the Soviet planning system promoted the development of tourism and the establishment of a tourist infrastructure; the area became quickly famous in the region for recreation and leisure.

The Soviet planning institutions justified the ecological relevancy of the walnut-fruit forests by their positive impact on the water cycle necessary for the irrigation system in the Fergana Valley and by the great species variety which could be used for the development of new hybrids and thus for the improvement of nut and fruit plantations elsewhere in the USSR (Gosudarstvennyi Komitet SSSR po lesu 1990 – 1991: 71). In other words, the ecological value of the forests was primarily justified economically, and thus forest protection was considered necessary for economic reasons.

Apart from the official economic value of the forests, the autochthonous population also valued the forests for spiritual reasons: Several sacred places are located in the forests to which people from the wider region pilgrimaged to pray for healing of their ailments or for fulfilment of their desire for children. The party officials tried to stop such pilgrimages and saw these popular beliefs as a serious problem in spite of the atheistic Soviet ideology (ESHIMBETOV 1962).

Since the central command economy of the USSR prioritized cotton cultivation against other agricultural or forestry efforts in Central Asia people of the whole area, including the walnut-fruit forest region, were forced to work on cotton fields in the Fergana valley. Gathering nuts and fruits was subordinate and brought little economic gain for the collectors. The forced assignments on the cotton fields, the centralistic management and decision finding without consultation and participation of the local population, and the prohibition of practising Islam led to a further alienation between the local population and the regional *nomenclatura*. Over several decades the leading positions in the local administration, party organs and *leskhoze* were filled with ethnic Russians or Europeans which resulted in a feeling of inferiority on the part of the local ethnic Kyrgyz and Uzbeks.

To summarise, various political developments and governmental policies influenced significantly the management and utilisation of the natural resources and thus the immediate interrelation between place-based actors and the environment. Obviously, decisions concerning the natural resources in Kyrgyzstan were reached at national levels and implemented in command style, leading to indifference of the local people with regard to a sustainable resource management. The collapse of the Soviet Union and the independence of the Kyrgyz Republic in 1991, interlinked with tremendous economic and social ruptures, have brought about new fundamental changes in the management and utilisation intensity of land and forest resources. The present web of institutions, actors and their concerns related to the walnut-fruit forests and the surrounding lands will be analysed in the following.

The Web of Actors, Interests and Institutions in Kyrgyzstan's Walnut-Fruit Forests

At first glance, the explanation of the present degradation of the walnut-fruit forests in Kyrgyzstan might be simple: There are just too many people overusing or misusing the forests and other land resources by cutting trees, bushes and shrubs to gain firewood or to extend their meadows or fields, herding their livestock in the forests and harming their rejuvenation, while the governmental forestry farms are not able to control their land or to afforest adequately. It is self-evident that such an essentially Neo-Malthusianian explanation is insufficient and ignores historical and non-place-based developments. Undoubtedly, population numbers in the area have increased significantly over the past decades: the numbers tripled from 1959 till today (National Statistical Committee of the Kyrgyz Republic 2001). But the interdependencies of people living in the area with surrounding land and forest resources, the lack of alternative means for income generation, and the causes of high population numbers are not given facts at all. For an understanding of the present environmental change it is necessary to identify place-based and non-place-based actors, their concerns, and the governing institutions against the historical background.

As was shown in the previous chapter, the walnut-fruit forests and their surroundings offer various land and forest resources which resulted in an elaborate utilisation system created over the past one hundred years. In the Soviet era more or less all employable inhabitants of the area found employment in state-run enterprises or governmental institutions. After the collapse of the USSR many of these enterprises closed down or reduced their workforce dramatically when the state budgets were tightened, so that many people lost their jobs.

With regard to the ratio between the market economic value of the natural resources and population numbers, it becomes obvious that the forestry sector was highly subsidised in the Soviet era, so that the region is overpopulated in economic terms, in the sense that there are only limited employment alternatives irrespective of natural resources. Today, inhabitants of the area are no longer able to sustain their livelihood by paid employment in state enterprises. Consequently, they have had to change their livelihood strategies, in which the intensified utilisation of the nearby natural resources play a major role, including arable farming on small plots, animal husbandry, gardening, collection of firewood, nuts and fruits to meet subsistence needs or to generate income. Obviously, the concerns and in consequence the impact of local actors on natural resources have changed significantly since independence. Prior to independence they used the resources only to a limited degree because they were not dependent on high yields. Today, arable, pasture and forest resources play a much more prominent role in their livelihood strategies. According to my investigations, members of almost all households in the relevant villages keep livestock, collect walnuts and firewood, and a high percentage collect fruits (90 %), morels (60 %) and herbs (45 %) (Schmidt 2005b: 101). The private herds graze on high pastures during the summer, but in the clear forests in spring and autumn. At the end of the summer farmers cut grass in the forests to gain winter fodder for the increased droves. Cattle and sheep are a popular investment because livestock keeping is a profitable business and animals are flexible capital because they can easily be transformed into cash when needed.

In general, the web of actors and concerns has changed substantially over the last 15 years. Besides the intensified usage by the locals due to pressing needs there is pressure on land and forest resources by new external actors, too. Within the frame of economic liberalisation in Kyrgyzstan since independence (Dabrowski & Antczak 1995; Abazov 1999; Dana 2000), mainly foreign private merchants and companies stepped in to carry out trading businesses. The walnut business is nowadays in the hands of mainly Turkish companies, though they employ people from the nearby towns to open, sort and pack the collected nuts, which afterwards are exported to Turkey or the Gulf States. Wild apples are today processed to concentrate by a Chinese enterprise that opened in Jalalabad in 2002; the apple concentrate is exported to China. Although cutting and selling of timber is officially limited to the state-run *leskhozes*, there is a large unofficial market. Several private enterprises process timber from the forests, and foreign wood companies are interested in burl and root wood of nut trees. Many old trees were cut for this purpose during the last 16 years and the valuable wood, which is used for exquisite veneers for items such as chessboards, gun butts or for the interiors of luxury cars, is exported to North America or Europe; a business that is officially prohibited but which provides officials on all levels of the administration with their share. A similar change of demands concerns morels: in the past, they were collected only for private consumption, whereas they are now highly esteemed by global demand. Some salesman export dried morels to France and Japan where they are sold as delicacies at high prices.

Besides the above-mentioned economic interests in the forests political concerns are also prevalent. The government, self-evidently, tries to control the territory and thus keeps the right of ownership in its hand. However, there are quarrels between different administration levels about the competency for specific lands. In contrast to other areas of Kyrgyzstan, the local administrations (*ailökmötü*) are only responsible for the settled areas, whereas arable lands are still affiliated to the *leskhozes,* and high pastures to the *rayon* (county). The local councils claim – as yet without success – that these lands should be transferred to them because land is an important source of income.

Scenery, fresh air and pleasant summer temperatures are valuable factors for tourism und thus for the interests of a group of other actors: managers and employees at local and non-local levels of private, governmental or international tourist enterprises as well as the tourists themselves have become relevant actors with regard to resource management and environmental change (cf. Kirchmayer & Schmidt 2005). The construction of resorts and the leisure activities of tourists stand in competition with agricultural, forestry or conservation aims.

Specific discourses on Kyrgyzstan's environment must also be considered. The ecological role of the forests is still a prevalent meta-narrative, and the state feels responsibility for the protection of these forests. However, the governmental role is ambivalent because the forest service realises economic profits from the forests and is involved in semi-legal wood cutting, but declares reservation zones without having adequate means to implement the necessary measures. National and international scientists contribute to a prolongation and accentuation of the environmental narrative by pointing out the uniqueness of these forests within their research efforts (cf. BLASER et al. 1998; SUCCOW 2004).

As regards institutional settings, the present situation shows a considerable lack of effective and accepted rules. Although several organs such as local government, forest farms, and councils of elders with specific competencies do exist formally, the fairness, implementation and acceptance of rules by those concerned are questionable. Specific usufruct rights are based only on vague permission. For instance, the right to cut grass on specific plots given – mainly orally – in the Soviet period still prevails, whereas the right to harvest nuts is given to local households on a yearly basis only. Quarrels and irregularities are common when usufruct rights for nut collection are allocated, especially in places with high numbers of inhabitants and limited forest resources. In some villages, households were given no more than eight to ten walnut trees which are not necessarily located on the plot on which the household has permission to cut grass. The nuts on a specific territory in the forest can thus be harvested by one household, while members of another household cut grass and again others collect morels or apples; the *leskhoze* takes timber and firewood out of the same plot. The situation becomes even more confusing when stakeholders transform their rented forested plot into hay meadows or arable land (cf. MESSERLI 2002).

Generally, power relations are asymmetric. Quarrels about competencies between local administration and *leskhoze* as well as institutional weakness hinder the development of a sound and widely accepted strategy of resource utilisation. Local inhabitants have no trust in official institutions in which corruption and nepotism prevail, while other relationship ties are more effective. Local actors with connections to persons in key positions have more agency options than people without such networks. Influential non-place based actors such as foreign wood companies or members of the state forest service can achieve their goals relatively easily because economic hardships as well as the vagueness and weakness of institutions do not hinder them. Although local stakeholders are interested in long-term sustainability of resource use, their intensive use and even overuse of the nearby forests become understandable in view of their present economic needs. They see their surrounding environment as an agricultural resource to sustain their livelihoods rather than sharing the prevailing opinion by Western scientists or governmental ideals that the walnut-fruit forests are a fragile entity threatened by intense usage. Nevertheless, other external actors are much more profit-orientated and devoted to extracting resources than to preserving environments. The link between environmental degradation and weak institutions becomes obvious.

Conclusion

My remarks on institutions, actors and concerns in Kyrgyzstan's mountain resources are intended to show that mountains are not peripheral and isolated areas without history but linked by chains of influences, dependencies and concerns of actors and institutions on various scales with the wider world. Mountain environments contain valuable resources and thus become arenas of conflicting actors and concerns. Hence, the question of resource utilisation is a question of power within a frame of institutional regulations.

Practising political ecology in Post-soviet Mountain spaces carries the responsibility of engaging with Tsarist colonialism and Soviet modernisation processes because present human-environmental relations in Kyrgyzstan cannot be understood outside of or apart from historical experience. The legacy of the Tsarist and Soviet systems of resource allocation and utilisation is still prevalent and influences present management strategies to a major degree. Recent developments and forces of globalisation processes alike have a significant impact on local agencies, as the conflicting aims between local population and international companies show. Owing to its historical and spatial dimensions of analysis, political ecology is a feasible approach to analyse the complicated diversity of actors, concerns and institutions in mountain areas.

References

ABAZOV, R. (1999): Policy of economic transition in Kyrygzstan. Central Asian Survey 18 (2): 197-223.

ADGER, W. N. (2000): Social and ecological resilience. Are they related? Progress in Human Geography 24 (3): 347–364.

ALLAN, N. J. R. (1986): Accessibility and altitudinal zonation models of mountains. Mountain Research and Development 6 (3): 185–194.

BERKES, F. (2007): Understanding uncertainty and reducing vulnerability. Lessons from resilience thinking. Natural Hazards 41: 283–295.

BERKES, F.; J. COLDING; C. FOLKE (2003): Navigating social-ecological systems. Building resilience for complexity and change. Cambridge.

BHASKAR, R. (1986): Scientific realism and human emancipation. London.

BLAIKIE, P. M.; H. C. BROOKFIELD (1987): Land degradation and society. London.

BLAIKIE, P. M. (1985): The political ecology of soil erosion in developing countries. London.

BLAIKIE, P. M. (1995): Changing environments or changing views? A political ecology for developing countries. Geography 80 (3): 203–214.

BLASER J., J. CARTER; D. GILMOUR (eds.) (1998): Biodiversity and sustainable use of Kyrgyzstan's walnut-fruit forests. IUCN, Gland and Cambridge, and INTERCOOPERATION, Bern.

BRYANT, R. L. (1998): Power, knowledge and political ecology in the Third World. A review. Progress in Physical Geography 22 (1): 79–94.

BRYANT, R. L.; S. BAILEY (1997): Third World political ecology. London, New York

DABROWSKI, M.; R. ANTCZAK (1995): Economic reforms in Kyrygzstan. Russian & East European Finance & Trade 31 (6): 5–30.

DANA, L. P. (2000): Change and circumstances in Kyrygyz markets. Qualitative Market Research: An International Journal 3 (2): 62–73.

Department for International Development (DFID) (2000): Sustainable livelihoods guidance sheets. London.

DIKAU, R.; KREUTZMANN, H. & M. WINIGER (2002): Zwischen Alpen, Anden und Himalaya. In: EHLERS, E. & H. LESER (eds.): Geographie heute – für die Welt von morgen. Gotha. 82–89.

Direktor Lesnogo Departamenta (1902): Lesnoe delo v Turkestane. (Iz Otcheta Direktora Lesnogo Departamenta po poezdk v 1900 godu v Turkestanskii krai. (= Forests of Turkestan. From the report of the Director of Forest Department on his journey to Turkestan in 1900). Lesnoi Journal 6: 431–472.

Distanova, V. (1974): Istoriia leskhoza imeni Kirova Leninskogo raiona Oshskoi oblasti. (= History of Kirov leskhoze of Lenin rayon, Osh oblast.) Diploma thesis at the Faculty of History of the Kyrgyz State University. Frunze.

Eckholm, E. (1975): The deterioration of mountain environments. Science 189:764–770.

Ehlers, E.; H. Kreutzmann (eds.) (2000): High mountain pastoralism in Northern Pakistan. Erdkundliches Wissen 132. Stuttgart.

Escobar, A. (1996): Constructing nature. Elements for a post-structural political ecology. In: Peet, R.; M. Watts (eds.): Liberation ecologies: environment, development, social movements. New York. 46–68.

Eshimbetov, T. T. (1962): O merakh po prekrashcheniiu plomnichestva k tak nazyvaemym 'sviatym' mestam v Bazar-Kurganskom i Ala-Bukinskom raionakh. KP Kirgizii, 16 Iiunija.

Forsyth, T. (1998): Mountain myths revisited. Integrating natural and social environmental science. Mountain Research and Development 18 (2): 107–116.

Forsyth, T. (2003): Critical political ecology. The politics of environmental science. London.

Gerrard, J. (1991): Mountains under pressure. Scottish Geographical Magazine 107 (1): 75–83.

Gosudarstvennyi Komitet SSSR po lesu (1990–1991): Proekt organizacii i razvitiia lesnogo khoziaistva Kirovskogo leskhoza. Tom 1. Moskva.

Gottschling, H.; I. Amatov; G. Lazkov (2005): Zur Ökologie und Flora der Walnuss-Wildobst-Wälder in Süd-Kirgisistan. Archiv für Naturschutz und Landschaftsforschung 44. Greifswald. 85–130.

Gosudarstvennyi oblastnyi archiv Jalalabad: Lesnoi zakon 1921 (f. 806, op.1, d.32 i. 4). To the Council of People's Commissars, 5 April 1918 (f.126, op.1, d.362)

Decree N°7136-R by the Council of People's Commissars of the USSR, 30 April 1945 (f.76, op.1, d.18, l.14)

Grötzbach, E. (1982): Das Hochgebirge als menschlicher Lebensraum. Eichstätter Hochschulreden 33. München.

Hartwig, J. (2007): Die Vermarktung der Taiga: Die Politische Ökologie der Nutzung von Nicht-Holz-Waldprodukten und Bodenschätzen in der Mongolei. Erdkundliches Wissen 143. Stuttgart.

Hewitt, K. (1988): The study of mountain lands and peoples. In: Allan, N. J. R.; G. W. Knapp; C. Stadel (eds.): Human impact on mountains. Boston. 6–23.

Hewitt, K. (1992): Mountain hazards. GeoJournal 27 (1):47–60.

Huber, U. M.; H. K. M. Bugmann & M. A. Reasoner [eds.] (2005): Global change and mountain regions. An overview of current knowledge. Dordrecht.

Ives, J. D.; B. Messerli (1984): Stability and instability of mountain ecosystems: lessons learned and recommendation for the future. Mountain Research and Development 4 (1): 63–71.

Jentsch, C. (1984): Für eine vergleichende Kulturgeographie der Hochgebirge. In: Grötzbach, E.; G. Rinschede (eds.): Beiträge zur vergleichenden Kulturgeographie der Hochgebirge. Eichstätter Beiträge 12. Eichstätt. 57–71.

Jodha, N. S. (1997): Mountain agriculture. In: Messerli, B.; J.D. Ives [eds.]: Mountains of the world: a global priority: a contribution to chapter 13 of Agenda 21. Ney York: 313–335.

Kirchmayer, C.; M. Schmidt (2005): Transformation des Tourismus in Kirgistan: Zwischen staatlich gelenkter *rekreacija* und neuem *backpacking*. Tourismus Journal 8 (3): 399–417.

Korzhinskii, S. (1896): Rastitel'nosti Turkestana I–III. Zakaspiiskaia Oblast', Fergana i Alai. (= Studies on plants in Turkestan. I – III. Zakaspiiskaia region, Fergana and Alai.) Sankt Peterburg.

Kreutzmann, H. (1993): Entwicklungstendenzen in den Hochgebirgsregionen des indischen Subkontinents. Die Erde 124 (1): 1–18.

Kreutzmann, H. (2001): Development indicators for mountain regions. Mountain Research and Development 21 (2): 34–41.

Krings, T.; B. Müller (2001): Politische Ökologie. Theoretische Leitlinien und aktuelle Forschungsfelder. In: Reuber, P.; G. Wolkersdorfer (eds.): Politische Geographie: handlungsorientierte Ansätze und Critical Geopolitics. Heidelberger Geographische Arbeiten 112. Heidelberg. 93–116.

Lisnevski, V. I. (1884): Gornye lesa Ferganskoi oblasti. Novyi Margelan.

Messerli, B.; J. D. Ives (eds.) (1997): Mountains of the world: a global priority. A contribution to chapter 13 of Agenda 21. New York.

Messerli, S. (2002): Agroforestry. A way forward to the sustainable management of the walnut fruit forests in Kyrgyzstan. Schweizerische Zeitschrift für Forstwesen 153 (10): 392–396.

Musuraliev T. M. (1998): Forest management and policy for the walnut-fruit forests of the Kyrgyz Republic. In: Blaser, J.; J. Carter; D. Gilmour (eds.): Biodiversity and sustainable use of Kyrgyzstan´s walnut-fruit forests. IUCN, Gland and Cambridge, and INTERCOOPERATION, Bern. 3–17.

National Statistical Committee of the Kyrgyz Republic (2001): Regions of Kyrgyzstan. Jalal-Abad Oblast. Results of the first national population census of the Kyrgyz Republic of 1999. Publication III (series R). Bishkek.

Navrockii, S. (1900): Materialy dlia lesnoi statistiki Turkestanskogo kraia. Lesnyia dachi Turkestanskogo kraia. Tashkent.

Neumann, R. P. (2005): Making political ecology. New York.

Nightingale, A. (2003): Nature–society and development. Social, cultural and ecological change in Nepal. Geoforum 34: 525–540.

North, D. (1990): Institutional change and economic performance. Cambridge.

Parish, R. (2002): Mountain environments. Harlow.

Peet, R.; M. Watts (1996): Liberation ecologies. Environment, development, social movements. New York.

Price, L. W. (1981): Mountains and man. A study of process and environment. Berkeley.

Price, M. (1999): Global change in the mountains. New York.

Rathjens, C. (1982): Geographie des Hochgebirges. 1 Der Naturraum. Stuttgart.

Rathjens, C. (1988): German geographical research in the high mountains of the world. In: Wirth, E. (ed.): German geographical research overseas. A report to the International Geographical Union. Tübingen.

Rauner, S. Iu. (1901): Gornie lesa Turkestana i znachenie ikh dlia vodnogo khoziaistva kraia. (= Turkestan's mountain forests and their impact on regional water management). Saint Petersburg.

Robbins, P. (2004): Political ecology. A critical introduction. Malden.

Sarmiento, F. O. (2000): Breaking mountain paradigms. Ecological effects on human impacts in mangaged Tropandean landscapes. Ambio 29 (7): 423–431.

Sayer, A. (1992): Method in social science. A realist approach. London.

Schlager, E.; E. Ostrom (1992): Property-rights regimes and natural resources. A conceptual analysis. Land Economics 68 (3): 249–262.

Schmidt, M. (2005a): Kirgistans Walnusswälder in der Transformation: Politische Ökologie einer Naturressource. Europa Regional 13 (1): 27–37.

Schmidt, M. (2005b): Utilisation and management changes in South Kyrgyzstan's mountain forests. Journal of Mountain Sciences 2 (2): 91–104.

Schweizer, G. (1984): Zur Definition und zur Typisierung von Hochgebirgen aus der Sicht der Kulturgeographie. In: Grötzbach, E.; G. Rinschede (eds.): Beiträge zur vergleichenden Kulturgeographie der Hochgebirge. Eichstätter Beiträge 12. Eichstätt. 31–55.

Smethurst, D. (2000): Mountain geography. The Geographical Review 90 (1): 35–56.

Soffer, A. (1982): Mountain geography – a new approach. Mountain Research and Development 2 (4): 391–398.

Stone, P. B. (ed.) (1992): The state of the world's mountains. A global report. London.

Succow, M. (2004): Schutz der Naturlandschaften in Mittelasien. Geographische Rundschau 56 (10): 28–34.

Svod zakonov Rossiiskoi Imperii (1892): Tom vtoroi. Polozhenie ob upravlenii Turkestanskogo kraia. Sankt Peterburg.

Troll, C. (ed.) (1972): Geoecology of the high mountain systems of Eurasia. Erdwissenschaftliche Forschung 4. Wiesbaden.

Troll, C. (1975): Vergleichende Geographie der Hochgebirge der Erde in landschaftsökologischer Sicht. Geographische Rundschau 27: 185–198.

Uhlig, H. (1976): Bergbauern und Hirten im Himalaya: Höhenschichten und Staffelsysteme – ein Beitrag zur vergleichenden Kulturgeographie der Hochgebirge. Tagungsbericht und wissenschaftliche Abhandlungen des 40. Deutschen Geographentages Innsbruck. Wiesbaden. 549–586.

Uhlig, H.; W. Haffner (eds.) (1984): Zur Entwicklung der Vergleichenden Geographie der Hochgebirge. Wege der Forschung 223. Darmstadt.

UNESCO (1974): Working Group on Project 6: Impact of Human Activities on Mountain and Tundra Ecosystems. Final Report, Man and Biosphere Programme, No.14. Paris.

Watts, M.; H.-G. Bohle (1993): The space of vulnerability. The causal structure of hunger and famine. Progress in Human Geography 17 (1): 43–67.

Winiger, M. (1992): Gebirge und Hochgebirge: Forschungsentwicklung und -perspektiven. Geographische Rundschau 44 (7–8): 400–407.

Zimmerer, K. (1993): Soil erosion and social (dis)course in Cochabamba, Bolivia. Economic Geography 69 (3): 312–327.

Zimmerer, K. S.; T. J. Bassett (eds.) (2003): Political ecology. An integrative approach to geography and environment-development studies. New York, London.

Boundary-Making and Geopolitical Diversity in the Pamirian Knot

Hermann Kreutzmann[1]

1 Introduction

Boundaries and diversity are used as distinguishing classifiers in numerous contexts. Boundary-making plays a major role in the political sphere when the momentum and range of power are at stake. Long-lasting effects can be observed in colonial contexts, in the process of nation-building and in the delineation of spheres of influence. Geopolitics has received a bad reputation when ideologies and dictatorial regimes aimed at the extension of their dominions. In the Central Asian context geopolitics have played a major role for socioeconomic development in the arena between different spheres of influence. The specific interests of superpowers of their time had long-reaching effects into the spatial and economic periphery. Exogenously stimulated developments often resulted in transforming local living conditions. When discussing the significance of colonial intervention and geopolitical interference we have to keep in mind external strategies and their implementation versus regional and local responses. The present-day perception of Afghanistan, Tajikistan, Kyrgyzstan and Pakistan as nation states is strongly linked to their political affiliation in the 20th century in general and during the Cold War in particular. Kirghiz as a Turkic language and Tajik, an Iranian language, are written in Cyrillic script, while the Tajik of Afghanistan, Dari, is written in the Arabo-Persian script, in a similar manner as Urdu in Pakistan. Tajik and Dari symbolize the difference in script, lexemes and loan words which symbolize the socio-political backgrounds in two languages which originate from Persian. The same applies for a number of minority languages spoken in the Pamirs and the Hindukush mountains. The Turkic idiom of Uigur experienced a shift from a Persian script towards Latin and back. In each instance a political move was involved. Presently in some countries the changes of scripts are discussed again as a symbol for independence, traditional values and breaking with colonial and geopolitical legacies.

2 Ecological Diversity and Spatial Utilization Patterns in Central Asia

On the macro scale Central Asia has been the sparsely settled periphery between Europe and Asia. Environmentally the region is characterized by steppe, desert and mountains with arid conditions in the lowlands and increasing precipitation and humidity with altitude resulting in snow-covered mountains, glaciation, high mountain pastures and scanty forests. Given these assets common utilization patterns of ecological resources are related to a bi-polar approach: extensive nomadism in the vast desert and steppe regions covering substantial areas with sparse vegetation cover. Animal husbandry as a prime strategy is enhanced by certain forms of mountain nomadism in the Hindukush, Pamirs and Tien Shan.[2] In contrast agriculture is limited to oases in which intensive

---

[1] Freie Universität Berlin, Institut für Geographische Wissenschaften; D-12249 Berlin, hkreutzm@geog.fu-berlin.de

[2] The specific utilization pattern of high mountain pastures - such as observed in the „pamirs"

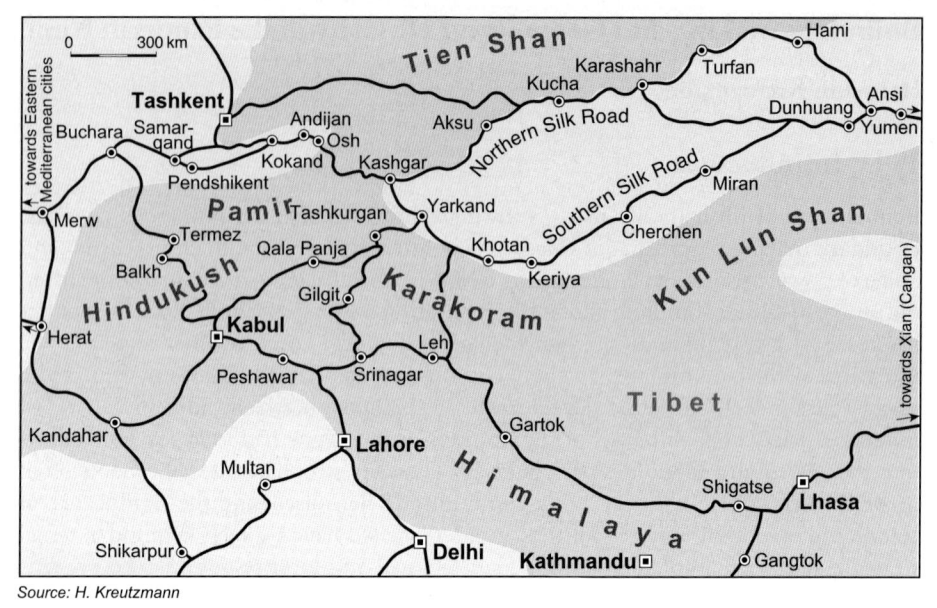

Fig. 1   Silk road network

Source: H. Kreutzmann

crop cultivation is linked to the demands of the bazaar towns and their surroundings along the traditional trade routes of the Silk Road network (Fig. 1). More important than silk has been cotton cultivation in major irrigated oases. Hydraulic resources for irrigation originate mainly from the glacier-fed rivers such as the Amu and Syr Darya issuing from the high mountain ranges within the desert-steppe environment. In the remoter mountain regions we find different forms of combined mountain agriculture (EHLERS & KREUTZMANN 2000: 15) in scattered mountain oases mainly supplied by gravity-fed irrigation schemes tapped from the tributary valleys of the main rivers. Niche production of valuable and marketable crops augmented the general pattern of grain crop cultivation for basic sustenance.

Economically and politically there existed competition between nomads and farmers over natural resources during long periods. While they competed in the production sector, political influence was mainly felt and contested in the urban centres of the oases towns. They were the prime target of all kinds of conquerors from Iran, Mongolia and China.[3] These historical events left their marks on the transforming Central Asian socio-economic landscape and prove the existence of a Eurasian exchange system over long periods of time.[4]

    (cf. KREUTZMANN 2003) – is characteristic for Central Asia and has repeatedly given scope for speculation about the economic potential of animal husbandry since Marco Polo's travels.

[3]   Cf. BREGEL 2003, CHRISTIAN 2000, KREUTZMANN 1997, 2002, 2004.

[4]   This led Andre Gunder Frank and Barry Gills to postulate a 5000 year-old Eurasian exchange network which in their opinion was instrumental for the development of China and Europe (GILLS & FRANK 1991). Consequently both authors observed an early emergence of a "world system" in Central Asia.

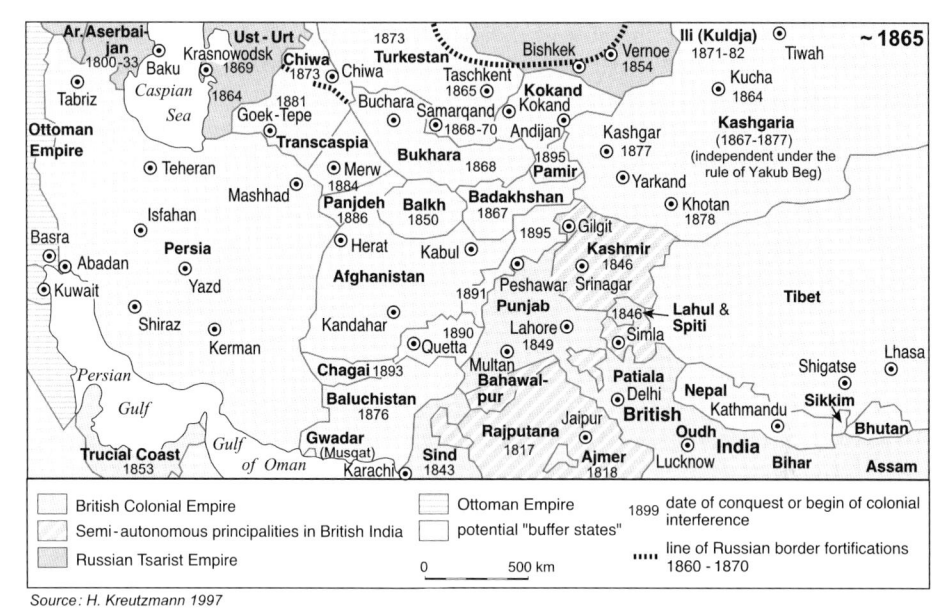

Fig. 2a Great Game I

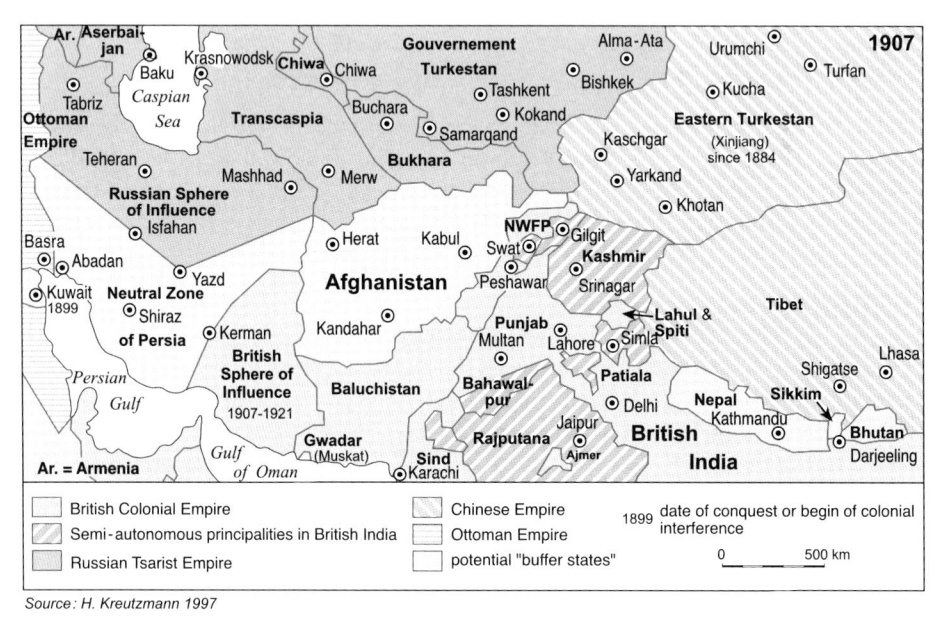

Fig. 2b Great Game II

In the 19th century its role changed significantly and the "Great Game" identified a polarisation that was stimulated by the prime interests of the two superpowers at that time (Fig. 2). Direct influence in the form of boundary-making and economic exploitation removed the former pattern of indirect control and tax-taking in a feudal system.

For the understanding of the present transformation process in Central Asia and the performance of independent states the geopolitical dimension of the "Great Game" and subsequent territorial demarcations needs to be discussed in greater detail.

3   The "Great Game" in Central Asia

In 1877 when Queen Victoria became the Empress of India Russian diplomats and military strategists debated about the importance of Central Asia from a Russian perspective, Colonel M. J. Veniukov vindicated "[…] the gradual movement of Russia in Central Asia [...]" as "[…] the re-establishment of extension of the sway of the Aryan race over countries which for a long period were subject to peoples of Turk and Mongol extraction".[5] Veniukov advocates a diffusion theory in which he identifies "[…] the mountainous countries at the sources of the Indus and the Oxus to be the cradle of the Aryan or Indo-European race. From this birthplace our ancestors spread far and wide […]."[6] After some deliberations abot the spread of people Veniukov concludes that the Russian advances in Central Asia can be interpreted as

> "[…] this 'return' of part of the Slavs to the neighbourhood of their prehistoric home […] We are not Englishmen, who in India do their utmost to avoid mingling with the natives, and who moreover, sooner or later, may pay for it by the loss of that country, where they have no ties of race […]. It is desirable that this historical result should not be forgotten also in the future, especially on our arrival at the sources of the Oxus, where we must create an entirely Russian border-country as the sole guarantee of stability of our position in Turkestan."[7]

The distinctive interests and justifications for the advance of both dominating powers contain strategies for "remigration" into an ancestral homeland and the exploitative interest in the wealth of Asia. Consequently, two types of colonies were created: Russian settlement colonies and British colonies of extraction. Nevertheless, the commonalities of both powers were discussed as well:

> "Possibly time will produce a radical change in the sentiments of the English, and then both great European nations will advance to meet the other in Asia, not with bitter suspicions and reproaches, but with confidence and benevolence as workers in the same historical mission – the civilization of the Far East. But will that time come soon? Russia, in any case, without awaiting it, must complete _her_ mission: the occupation of the whole of Turkestan. This, unquestionably, will prove not disadvantageous in that respect as well, that it will force England to be more on her guard in other lines of universal policy, in which she is ever antagonistic to the views of Russia."[8]

---

5   Political and Secret Department Memoranda: The Progress of Russia in Central Asia by Colonel M. J. Veniukoff (translated from the "Sbornik Gosudarstvennikh Znanii" 1877 (= IOL/P&S/18/C 17: 1).
6   Political and Secret Department Memoranda (= IOL/P&S/18/C 17: 1).
7   Political and Secret Department Memoranda (= IOL/P&S/18/C 17: 2).
8   Political and Secret Department Memoranda (= IOL/P&S/18/C 17: 22).

That envisaged time was not far away at the turn from the 19th to the 20th century. The British Viceroy in India, Lord Curzon, identified the Central Asian countries and territories in his famous statement as "pawns on a chessboard". British India and Russia were the players who gambled about their influence in Transcaspia, Transoxania, Persia, and Afghanistan (cf. Fig. 2). But this battle was not solely about regional control, it was a contest about the world domination of imperial powers. Great Britain had achieved already maritime supremacy, now the last land-locked area – Central Asia – came into focus. From a British viewpoint Central Asia posed the "buffer region" to protect more than pawns: the "jewel of the crown", a synonym for their possessions in India. From a Russian perspective expansion towards the East and the South was a consequential endeavour ever since Peter the Great had mentioned in his testimony that Russia's future was linked to Asia (cf. HAUNER 1989). Both superpowers expected sufficient wealth to be exploited from Central Asia to pay for their exploring adventures and military expenditure.

During the 19th century both superpowers reached a state of confrontation over contested supremacy in Central Asia. Both had literary celebrities justifying their cause and in both countries contemporary bourgeois debate highlighted the civilizing mission to be accomplished. Great Britain had Rudyard Kipling who was one of the foremost advocates of the "Great Game" and had coined the term of a "white man's burden" (cf. KREUTZMANN 1997). With missionary zeal and state authorization, civil society measures were to be promoted in Asia grounded in European standards. His Russian counterpart was Fjodor M. Dostojevsky who published an essay on the importance of Asia for Russia's future in which he justified the Asian conquest as a mission for the promotion of civilization. Dostojevsky compared the colonial expansion into Central Asia with the European conquest of North America (cf. HAUNER 1989, 1992). The second half of the 19th century experienced a heated debate in political and academic circles about the effects of the Anglo-Russian rivalry in Central Asia.

In Russia the Gorchakov Memorandum of 1865 marks the beginning of the animated phase of the "Great Game" (cf. Fig. 2a). The British Premier Disraeli responded in his famous speech at Crystal Palace 1872 in which he announced the imperial policies for further expansionism. Immediate results were the "forward policy" in the Afghan borderlands and the subsequent crowning of Queen Victoria as Empress of India (1877). Russia and Great Britain fought this game in the remote mountains of the Hindukush, Karakoram and Pamirs where their spies-cum-explorers met in unexpected locations. At the same time there was competition among the diplomatic staff posted in Central Asian centres. Notably Kashgar became one of the hotspots of confrontation where a weak Chinese administration personified by a Taotai fell prey to the powerful representatives of the superpowers: the Russian Consul M. Petrovsky and his British counterpart George Macartney were the protagonists and reported to their respective governments in detailed reports which give us historical evidence on the socio-economic conditions in Central Asia besides strategic and military intelligence during their rivalry.

The "Great Game" in its narrow definition came to an end in 1907 without any military encounter and no loss of lives. Russia and Great Britain came to terms and consented to the text of the so-called Anglo-Russian Convention in which respective spheres of influence, buffer states and regions of non-interference were agreed upon (cf. Fig. 2b). Instrumental for the accord was the "heartland theory" which drew geopolitical significance towards Central Asia.

The geographer Halford Mackinder formulated his "heartland theory" in 1904 which became one of the most influential texts of the geopolitical debate until today. Mackinder drew prime attention towards Central Asia as he stated that the Tsarist regional dominance was linked to their equestrian tradition from nomadic Asian backgrounds. From the safe retreat of the Inner Asian steppe regions conquests had taken off towards Europe, Persia, India and China. He described the European civilization as the result of a secular battle against Asian invasions (MACKINDER 1904: 423). The naval predominance of Great Britain and imperial control of world trade had been modified through a shift in terrestrial traffic structures. The Russian railways were perceived as the successors of the equestrian mobile forces. Central Asia had become the arena of contest, the more as a Russian-German and/or a Sino-Japanese alliance could contribute to a shift of world affairs to the "heartland" of the Eurasian continent which he perceived as a "geographical pivot of history" (MACKINDER 1904: 436). He predicted the transformation of Central Asia from a steppe region with little economic power into a region of prime geostrategic importance. Culture and geography would contribute to the key region. Mackinder identified four adjacent regions encompassing the heartland of "pagan" Turan in the shape of a crescent and denominated by religious affiliations: Buddhism, Brahmanism, Islam and Christianity (MACKINDER 1904: 431).[9]

Similar ideas of a Central Asian „heartland" or a pivotal role stimulated Owen Lattimore's perceptions in his book „Pivot of Asia" (1950). Keeping the experiences of World War II in mind Lattimore drew a circle with a diameter of 1000 miles around Urumchi and identified Central Asia as a "whirlpool" stirred-up by "political currents flowing from China, Russia, India and the Middle East" (LATTIMORE 1950: 3). By following the same Central Asian-centred approach Milan Hauner shifted the centre in the 1980s to Kabul, drew a similar circle and identified a world of "even greater contrasts" which "touches upon the volatile and oil-rich region of the Middle East" (HAUNER 1989: 7). The last statement has remained valid through the dissolution of the Soviet Union, the Taliban rule in Afghanistan and in the aftermath of 9/11 and the Iraq crisis. The fact that Ahmed Rashid (2000) subtitled his book on the Taliban as "Islam, Oil and the New Great Game in Central Asia" is only one case in point for the reference to the "Great Game" connotation of contemporary geopolitical problems in the region.[10] The presence of American and Russian troops on airports and along borders in Central Asia proves the continuing geopolitical significance of the region and its linkage to contemporary crises zones.

What are the effects of certain lines of thought and resulting political actions on Central Asia and why do we still refer to the metaphor of a „Great Game" when discussing contemporary strategic interference and socio-economic transformations in geopolitical

---

[9] With the passage of time Mackinder modified his theory under the impression of events during the First and Second World Wars and influenced the thoughts of Karl Haushofer and other geopoliticians of his time.
[10] Cf. KREUTZMANN 1997, 2002, 2004, 2006, 2007, ROY 2000.

contexts. Boundary-making and its impact on nation-building, economic and political participation severely influenced socio-economic developments in the mountainous areas of Central Asia. Some cases in point need to be introduced for the understanding of the far-reaching consequences of imperial border delineations. First of all, the practical impact on trade relations and economic exchange need to be investigated.

4 Development of Central Asian Trade and Attenuated Exchange Relations in the Aftermath of the „Great Game"

In Central Asia the "Great Game" resulted in the demarcation of international boundaries separating the spheres of influence of the super powers of the time. Besides executing direct control and domination in the core areas of their empires, Great Britain and Russia had created buffer states at the periphery such as Persia and Afghanistan (Fig. 2b). In their negotiations they had excluded Kashgaria or Eastern Turkestan which nominally was under Chinese administration. Trade between South and Central Asia was affected by this constellation and a rivalry had developed since British commercial interests entered this sector in 1874 (cf. BOSSHARD 1929; DAVIS & HUTTENBACK 1987; KREUTZMANN 1998). Both super powers competed for dominance on the valuable markets in the urban oases of the Silk Road such as Kashgar and Yarkand. According to the theory of imperialism, the merchants of the industrializing countries tried to purchase raw materials such as cotton, pashmina wool and hashish while in exchange textiles and manufactured products were offered in the bazaars (cf. KREUTZMANN 1998). Russia had some advantage as access was easier. From the railhead at Andijan in the Ferghana Valley, which was linked to the Middle Asian Railway in 1899 the distance to Kashgar (554 km) could be covered in twelve marches via Osh, Irkeshtam, and Ulugchat by crossing only one major pass, Terek Dawan (3870 m). On the other hand trade caravans from British India had to follow either of three trans-montane passages – the Leh, Gilgit, and Chitral routes – which were much longer and more difficult.

The competition for the Central Asian markets has continued after the October Revolution which caused the closure of the Russian/Soviet Consulate in Kashgar from 1920-1925. This event affected the Soviet commerce with Kashgaria detrimentally while the British share soared. Overall trade significantly declined due to the disturbances in Chinese Turkestan after 1935 and later due to World War II and the Chinese Revolution. Central Asian trade became an important factor in cross-boundary relations affecting the economies in the regions traversed for a period of forty years. The total annual volume of Indo-Xinjiang commercial exchange surpassed the two million rupees level for most of the era between 1895 and 1934.

At the end of the 19th century George Macartney the British Consul-General in Kashgar had summarized the situation:

> "The demand for Russian goods is without doubt ever increasing. Cotton prints of Moscow manufacture, as cheap as they are varied and pretty, are very largely imported. The bazaars of every town are overstocked with them, as well as with a multitude of other articles, amongst the most important of which may be mentioned lamps, candles, soap, petroleum, honey, sugar, sweetmeats, porcelain cups, tumblers, enameled iron

plates, matches, knives and silks. These articles, with few exceptions, could, but for the competition, be supplied from India. But we have gradually had to relinquish our position in favour of Russia, until at last our trade has to confine itself chiefly to articles of which we are the sole producers and in which there is no competition."[11]

British interests in securing a substantial share in this commercial exchange governed their imperial designs and had an impact on the mountain societies involved. At the turn of the century Ladakh and Baltistan were dominated by the Maharaja of Kashmir, Gilgit had become an agency (re-established in 1889) under the joint administration of a British Political Agent and a Kashmiri Wazir-i-Wazarat. Principalities such as Hunza and Nager were affiliated after their defeat in the 1891 encounters, which were fought under the pretext of opening the Gilgit route for commercial purposes. At the same time the Mehtar of Chitral transferred his sovereignty in external affairs to a British Agent and was remunerated with an annual subsidy and a supply of arms.

This part of the region under study was controlled and de facto commercially incorporated in the British Indian exchange system. Trade with Afghanistan followed its own rules and became part of the special arrangements with the ruling Amir in Kabul. The major hiatus occurred in the aftermath of the October Revolution when a process of separation and isolation began. The economic relations of the Soviet-dominated Central Asian regions were re-directed and amplified towards Russia while at the same time international borders were sealed and became effective barriers for trade. This process took time and lasted until the mid-1930s. With growing alienation between the Soviet Empire and the Chinese-dominated part of Eastern Turkestan a nearly complete interruption of exchange relations between Tajikistan-Kyrgyzstan and Kashgaria came to a halt by 1930 (KREUTZMANN 1996: 179). The undercutting of bazaar prices through the provision of cheaper commodities of the same quality in kolchoz shops led to the termination of trade in this sector. Similar developments took effect on the Soviet border with Afghanistan during the 1930s:

> *"During the past few years, the effect of Soviet policy has been to restrict, in an increasing degree, traffic, excepting state-controlled trade, from Soviet Central Asia across the Afghan frontier on the River Oxus. [...] more European Russian officers have been appointed to ensure that the frontier is effectively closed"* (IOL/P&S/12/2275, dated 13.10.1939).

Border delineation and the establishment of different socio-political regimes had resulted in a collapse of trade and exchange in this Central Asian region which lasted for nearly 60 years until the end of the Cold War. With few exceptions traditional trade links and exchange routes were interrupted for two generations and are only reanimated at a slow pace.

---

[11] Report of George Macartney of 1st October 1898, quoted from Captain K. C. Packman, Consul-General at Kashgar 1937: Trade Report. In: India Office Library & Records: Departmental Papers: Political & Secret Internal Files & Collections 1931-1947: IOL/P&S/12/2354, p. 1.

5 Boundary-Making and its Effects for Divergent Regional Developments

A few cases in point from the turn of the century may illustrate how mountain regions have been involved in the demarcation of spheres of influence. The contenders of the "Great Game" in High Asia agreed to lay down boundaries in the comparatively sparsely populated regions of the Hindukush and Pamir. Sometimes these borders were described as natural frontiers, scientific boundaries and dialect borders. The Durand Line of 1893 separating Afghanistan from British India/Pakistan epitomizes such an effort and has continued to function as the symbol of colonial border delineation referred to as the "dividing line" (FELMY 1993). In order to safeguard the physical separation of two imperial opponents, international borders were outlined and Afghanistan was created as a buffer state (Fig. 3). Local livelihoods and regional interests were neglected and of secondary importance. The Pashtun settlement region was divided into two parts following an arbitrary line through the Hindukush ranges. The traditional migratory paths of seasonal nomads between the Central Afghanistan highlands and the Indus lowlands were intersected along the Hindukush passes. Numerous clashes between tribal groups and imperial troops in the borderlands characterized the political relations in the frontier that served as a buffer belt on the fringe of the empire (FRASER-TYTLER 1953). Now a special legal status has been assigned to these regions as they are administrated as Federally or Provincially Administered Tribal Areas (FATA or PATA). The movement of nomads (*powindah*) and their herds now depends on bilateral political relations and has been restricted, but has not ceased.

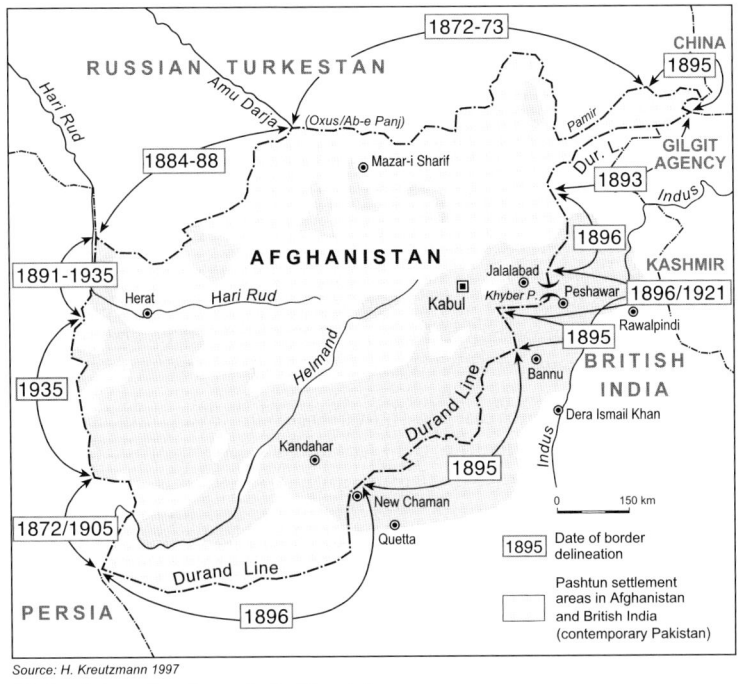

Fig. 3 **Afghanistan – borders of a buffer state**

For the British administrators the Durand Line served its purpose. Initially it was perceived as a "limit to the respective spheres of influence", but with the Treaty of Rawalpindi and subsequent agreements upto 1921 "the Durand Line, with certain modifications was declared to be the Indo-Afghan Frontier as accepted by the Afghan Government".[12] At least this was the official British position as recorded in the British Foreign Office. No division needs to last for eternity. Afghanistan and Pakistan inherited the Durand Line and the disputes about it. In recent years the Afghan government is challenging again the character of it. Some authors make a point that it was never accepted at all.[13] Nevertheless, diversity has increased on both sides of the Durand Line as communities have belonged over long stretches of time to different nation states, contrasting regimes, economies and societies. Presently it is unlikely that the dispute will fade away as the Durand Line is a volatile boundary in the efforts by the Pakistan Government and supporting international forces to wipe out Taliban strongholds there. For both governments the dispute opens chances to blame the other.

5.1   Wakhan, Shughnan and Darwaz – Symbols of Division in Badakhshan

Badakhshan came to the fore as a centrepiece of imperial interests. As Colonel Veniukov put it in 1877:

> *"The latter principality [Badakhshan] […] is unquestionably the most important of all those in Central Asia from a political point of view. Without possessing and colonising it we can never guarantee peace in Turkestan, or even the solidity of our rule there. It occupies the most flourishing district in the basin of the Oxus and feeds a numerous population. Possessed of it we could command the northern outliers of the Hindoo Kush and the passes over this range to the valley of the Kunar […] Unfortunately, since 1874 Badak[h]shan has been under the influence of the Emir of Cabul, although its inhabitants, Persians and Tadjiks, detest the Afghans […] without Badak[h]shan the Russians must consider themselves in Central Asia as guests, without any settled habitation and unable to form one. […] How history will solve the Badak[h]shan question of course cannot foreshadow; but it is impossible not to express admiration at the farsightedness of British policy."*[14]

As both contenders would not give up their claims for Badakhshan (Fig. 5a) the solution was negotiated to be division.[15] The remote corners of the Hindukush and Pamirs entered the limelight of imperial boundary-making, buffer-state creation and neutralizing corridors which separated respective spheres of influence.

The Wakhan Corridor of Northeastern Afghanistan symbolizes colonial border delineation. The southern limit is formed by the Durand Line (south) while the northern part came into existence as a result of the Pamir Boundary Commission of 1895 in which Russian and British officers negotiated the alignment, and Afghan officials assisted in

---

12   Both quotes are taken from a document titled "The Indo-Afghan Frontier" by the Research Department Foreign Office covering the Indo-Afghan relations between 1747 and 1947. In: IOL/P&S/12/1321: A survey of Anglo-Afghan Relations Part III: 26-27.
13   Cf. e. g. DJAN-ZIRAKYAR 1978; KHAN 1981; SCHETTER 2007: 241.
14   Political and Secret Department Memoranda (= IOL/P&S/18/C 17: 19).
15   For a detailed account of the process of division and the effects cf. KREUTZMANN 1996.

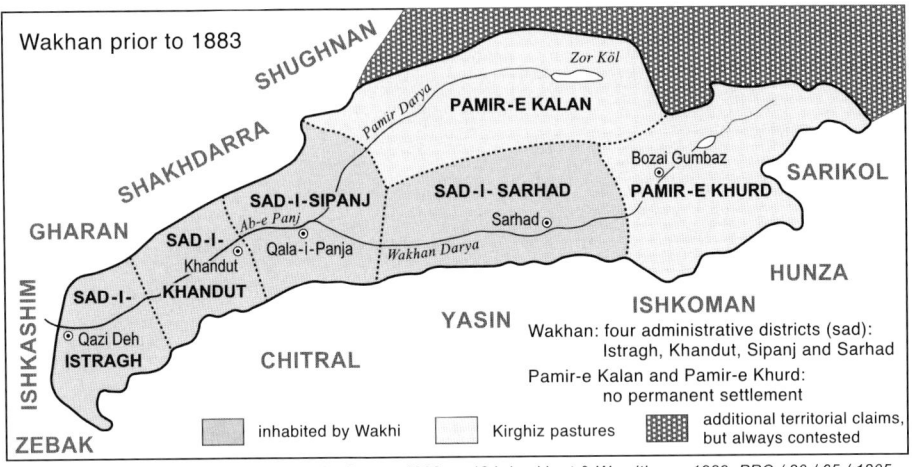

Fig. 4 Wakhan – a corridor separating the sphere of influence of two super powers

the demarcation (Fig. 4). This narrow 300 km-long and only 15–75 km wide strip was created to separate Russian and British spheres of influences and fulfilled the function to avoid direct military action between the two superpowers of that period and region. Part of the boundary follows the course of the Pyandsh (Amu Darya River), which was in accordance with the fashion of the time. The "stromstrich" boundary followed a role model tested in other regions of the world previously. The price for this colonial endeavour was the spatial partition of regional semi-autonomous principalities like Badakhshan, Darwaz, Wakhan, Shughnan, and Roshan (Fig. 5b). Subsequently both parts of each former principality experienced quite diverse socio-economic developments as part of greater political entities. Today we find regional units of the mentioned toponyms in Afghanistan and Tajikistan. The creation of these boundaries resulted in immediate refugee movements by ethnic minorities. Nevertheless, the effect of partition is felt in all areas, especially when international borders are closed and strictly controlled as it happened since the Cold War. Afghan Wakhan is suffering substantially from its dead-end location with missing through trade and exchange with neighbours (Felmy & Kreutzmann 2004). Nevertheless, the people living in the border areas have found practical solutions to border closures and are engaging in a local trans-boundary exchange of goods and services in the border triangle (Fig. 6).

Similar observations are valid for Shughnan and Roshan. In recent years relatives separated by a century-old border have re-established their relationship and the bridges across the Pyandsh River in Langar, Ishkashim and Khorog symbolize those endeavours. Here important symbolic changes have led to practical solutions to local needs. Bridges across the Amu Darya river have always played a vital role in cross-border affairs. The Langar and Ishkashim bridges were built to enable the Soviet army to invade Afghanistan (1979) and to safeguard their supplies from the Soviet Union for the control of Badakhshan. Meanwhile the function of the Ishkashim bridge has changed. For years during the war in Afghanistan support for the Northern Alliance and humanitarian aid for the suffering civilians were transported across this bridge.

Fig. 5a   Badakhshan in mid-19th century

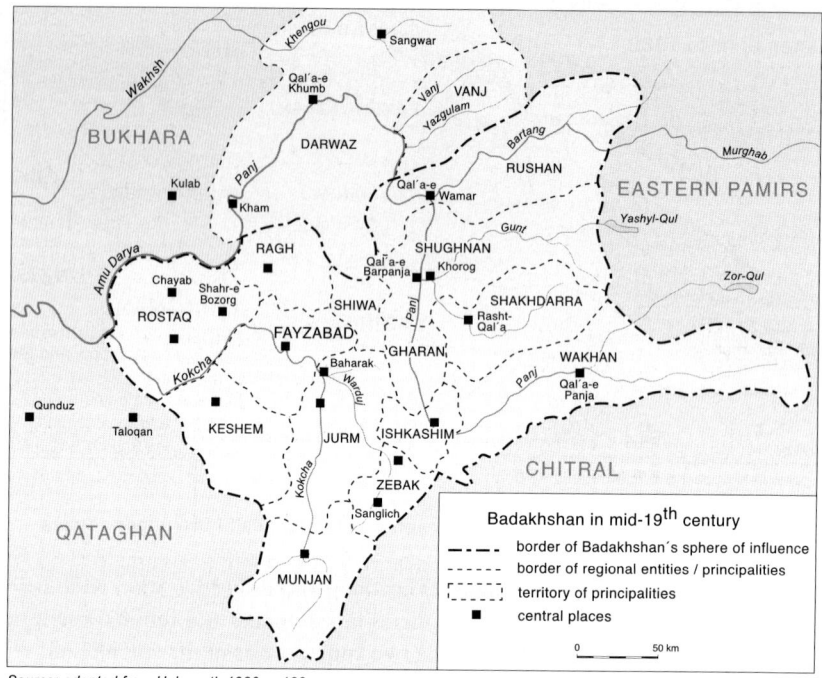

Source: adapted from Holzwarth 1980, p.188

Fig. 5b   Badakhshan – present administrative set-up

Source: own adaptation based on AIMS 2007, Hauser 2005

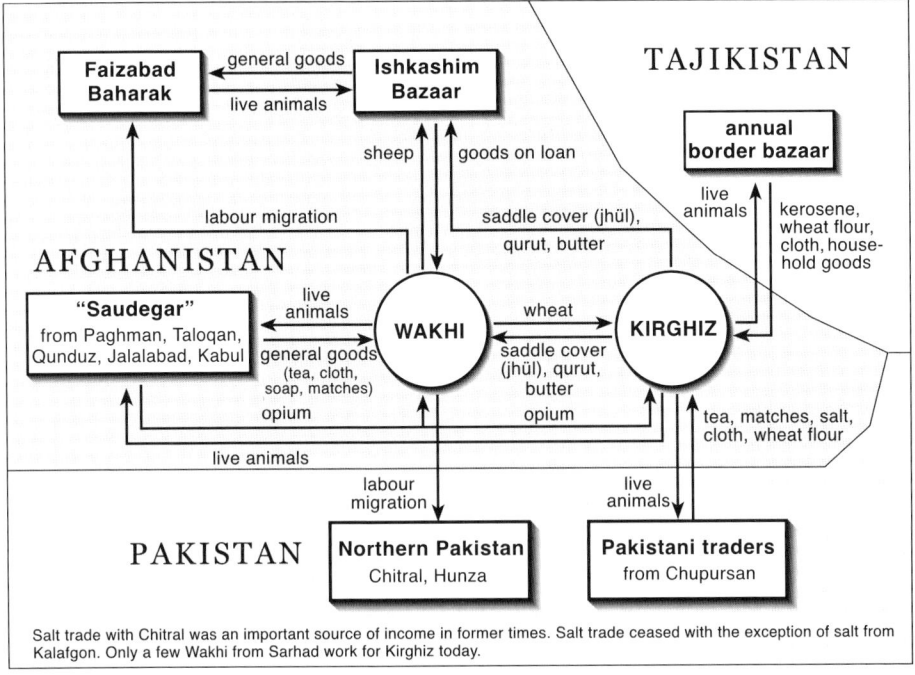

Fig. 6 Cross-boundary exchange relations of Kirghiz and Wakhi households

(Figure shows cross-boundary exchange relations between Tajikistan, Afghanistan, and Pakistan, with Wakhi and Kirghiz at the center. Key flows include: Faizabad/Baharak ↔ Ishkashim Bazaar (general goods, live animals); Ishkashim ↔ Wakhi (sheep, goods on loan, saddle cover (jhül), qurut, butter); annual border bazaar ↔ Kirghiz (live animals, kerosene, wheat flour, cloth, household goods); "Saudegar" from Paghman, Taloqan, Qunduz, Jalalabad, Kabul ↔ Wakhi (live animals, general goods (tea, cloth, soap, matches), opium); Wakhi ↔ Kirghiz (wheat, saddle cover (jhül), qurut, butter, opium); Kirghiz ↔ Pakistani traders from Chupursan (tea, matches, salt, cloth, wheat flour, live animals); Northern Pakistan (Chitral, Hunza) ← labour migration; live animals flows to Pakistan.)

Salt trade with Chitral was an important source of income in former times. Salt trade ceased with the exception of salt from Kalafgon. Only a few Wakhi from Sarhad work for Kirghiz today.

Source: own fieldwork 2003

The island in the river near Ishkashim became a storehouse for humanitarian aid such as wheat flour, milk powder and vegetable oil. The Tem bridge near Khorog was built by AKDN in order to link the cut-off Shughnan region of Afghanistan with Tajik Shughnan and to establish a market access. More severe was the dead-end location for Darwaz. The Ruzwai bridge near Kalai Khum (Fig. 7) enables peoples from Afghan Darwaz not only to participate in cross-border trade, but has terminated their exclusion from getting opportunities to augment stocks and supplies. Earlier on the people from Darwaz had to engage in a two-week journey on foot or by donkey to reach the bazaars of Faizabad. For the first time in 75 years people from Darwaz and Shughnan are now in a position to access markets all-year-round.

5.2 China's boundary with Afghanistan and Tajikistan

The assigned function of the Wakhan border was the separation of the British and Russian spheres of influences. The concept of buffer states was applied in China as well. Xinjiang and Tibet functioned somehow as buffer zones. The missing border link is the short Sino-Afghan boundary, which in itself is part of a disputed frontier. According to Chinese opinion, their border with Afghanistan and Tajikistan extends much further west. The ambiguity concerning Chines claims to the Pamirs did not escape the scrutiny of the Pamir Boundary Commission and of the British Consul-General in Kashgar, George Macartney, who himself went on an inspection mission towards the Little Pamir in 1895 and reported from Kizil Robat:

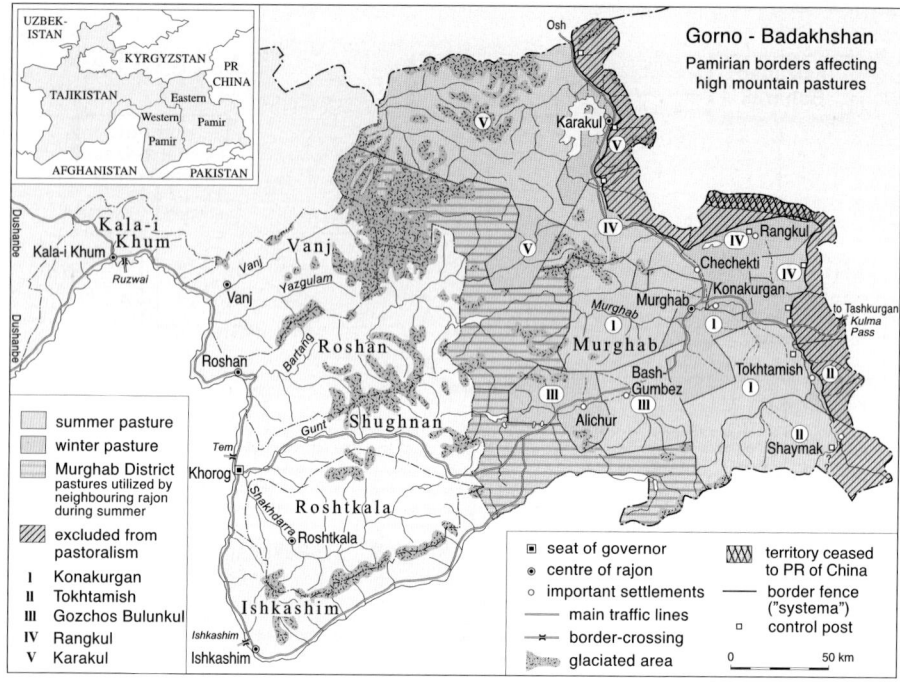

Fig. 7 Gorno-Badakhshan – a peculiar boundary sitution in the Afghan-China borderland

"From enquiries made by myself, it appears that, previous to that period, the Chinese jurisdiction extended westwards on the Alichur Pamir to Sumatsh and on the Great Pamir to the eastern end of Victoria Lake. The Khirgiz living in the Upper Oxus Basin within these limits and about Rangkul and Murghabi, owned a sort of loose allegiance to China not however as Chinese subjects, but rather as inhabitants of a State tributary to China. The Chinese appear to have never had much to do with the Small Pamir, that country having in times past been a dependency of Kanjut living in it subjected to Kanjuti taxation."[16]

George Macartneys observation highlights the fact that in remote areas of the Pamirs territorial control was less important than tributary relationships which could be well kept with more than one mighty neighbour. To claim territory based on these changing loyalties is as futile as is the notion of clear-cut boundaries at the time. Over time the boundaries have become visible and changed the fate of the abutters.

The factual contemporary boundary is agreed on by China's neighbours. Although claims to disputed territory differ substantially, China and Tajikistan are factually separated by a highly visible structure: the *sistema*, a barbed-wire fence which forms a present-day buffer of neutralized territory. The Kirghiz animal husbanders of this

[16] India Office Library & Records: Files relating to Indian states extracted from the Political and Secret Letters from India 1881-1911: Pamir Delimitation Commission 9. Oct. 1895 No. 195 (Reg. No. 451): IOL/P&S/7/82.

border region suffer from loss of accessible pasture due to the buffer arrangements (Fig. 7). Only by a written permission (*propusk*) they were allowed to cross into the neutral zone. Recently China and Tajikistan agreed that an area covering app. 976 km² should be handed over to the mighty neighbour. The affected herdsmen of the Rangkul community (*jamuat*) are more than disappointed about the loss and the prospect of no compensation for the expropriation of valuable pasture land. All these borders formed an integral part of the major global divide after World War II. The frontlines of the Cold War followed their historical predecessors. Western and Eastern alliances, as well as neutral states like Afghanistan (up to 1978) and the independent anti-Soviet path of Chinese communism (since 1958) met in the Pamirian knot. Thus, a remote mountain region became a meeting-point of competing political systems.

The alleviation of this confrontation did not terminate any military action in the region. The Pamir Boundary presently separates the newly independent state of Tajikistan (since 1991) from Afghanistan. The previous global confrontation has been replaced by regional conflicts. Nevertheless, these examples are not singular cases. Nearly all borders of the Hindukush-Himalayan arc are under dispute by one or the other side.

5.3 Border Disputes within the Soviet Union and thereafter

The attempt of Soviet nationalities' policies was to create new republics, which should represent the ethnic groups of Central Asia in adequate spatial and administrative settings. Consequently by 1929 ethnonymous republics were created to represent Kazakhs, Kirghiz, Tajiks, Uzbeks and Turkmens. The new republics did not have any boundaries in common with their predecessors, the Khanate of Khiva, the Khanate of Bukhara and the Turkestan Governorate-General. If the term "artificial boundaries" could be appropriate in any context, it would be here. The newly defined republics consisted of a spatial nucleus, but very often they had in addition satellite territories of enclaves and exclaves within the territory of neighbouring republics (Fig. 8). While this phenom-

Source: based on official maps of Tajikistan and Kyrgyzstan 1997, 1999

Fig. 8 Central Asian boundary disputes in the aftermath of the dissolution of the S. U.

enon did not pose grave differences during the period of the Soviet Union – basically all territories were under the central command of the Kremlin and only international boundaries with neighbouring countries such as China and Afghanistan were of any importance and hermetically sealed – another cause of germinated dissent erupted after independence in the early 1990s. Republican boundaries within the Soviet Union became international borders of sovereign states such as Uzbekistan, Tajikistan and Kyrgyzstan. In a survey two years after independence the Moscow Institute of Political Geography recorded 180 border and territorial disputes in the aftermath of the dissolution of the Soviet Union (HALBACH 1992: 5). Central Asia was no exception in this regard and these conflicts have increased since. According to a recent report of the International Crisis Group (*ICG* 2002) there is no Central Asian country without border disputes with its neighbours. To illustrate the scope of conditions and demands a few cases are listed: Irredentistic movements in Turkmenistan expect Uzbekistan to "return" the territory of the Khanates of Khiva and Khorezm. Tajik nationalists demand the "return" of Samarkand and Bukhara. Uzbekistan lays claim on the Eastern part of the Ferghana Valley, i.e. the Osh Oblast, the present-day economic and commercial centre of Southern Kyrgyzstan. The Uzbekistan government does not permit colleagues from neighbouring republics to consult the archival material in Tashkent, which documents the boundary decisions from the 1920s. Rental arrangements and the production of natural resources in exclaves from Soviet times are under dispute such as the Uzbek exploitation of oil and gas fields in Southern Kyrgyzstan and the deviation of irrigation water from the Andijan reservoir towards the Ferghana Valley (Fig. 8). The Ferghana Valley alone is containing seven enclaves through which major traffic routes are leading. The freedom of travel is more restricted than before as new measures of visa regulation of travel have been introduced. Some of these measures have been justified in the aftermath of attacks from Afghanistan-trained rebels, which plundered Tajik and Kirghiz villages on their way to the Ferghana Valley in 1999 and 2000. The future of rented lands and exclaves that were created for the protection of ethnic minorities is at stake and neighbouring governments discuss options for forced evacuation and migration to initiate population exchange.

5.4 Future Prospects and Conflict Resolution

The hope for friendly relations and mutual understanding has suffered several setbacks in recent years. All negotiating partners are interested in most favourable results from their national perspective. On a regional scale there is some hope since the Shanghai Cooperation Organization (SCO) which was founded in 1996 as the Shanghai-5 (Russia, PR of China, Kazakhstan, Kyrgyzstan, and Tajikistan) and became a fully-fledged organization under the name of SCO in 2001 when Uzbekistan joined.[17] The mandate is to improve mutual relations and to improve Central Asia's economic competitiveness in a globalized world. Therefore the SCO has supported the opening of new trade corridors between the PR of China and Kyrgyzstan (Irkeshtam Road) and Tajikistan (Khulma Road) respectively. The two major regional players – Russia and PR of China – cooperate with the European Union to link the Central Asian repub-

---

[17] The SCO became an internationally acknowledged organization in 2004 and operates a secretariat from Beijing. In August 2007 the most recent SCO meeting of political leaders was held in Bishkek.

lics through a road network (TRACECA route) with Europe via the Caucasus. The participation in regional and international trade may be one of the prime stimulants to overcome the legacies of previous geopolitical interference and reflect the economic interests of the big economic players of today in the future of Central Asia.

Nevertheless, the region under study suffered not only directly from Cold War confrontation but as well from regional problems, which remain to be a colonial legacy, but have developed into a conflict between neighbours. After more than 50 years of independence India and Pakistan are still engaged in military confrontation that is affecting economic exchange tremendously and keeps the mountain regions of the Karakoram and Western Himalaya in a state of dispute and uncertainty.

6   Effects of Boundary-Making and two Transformations in Central Asia

The starting point for our thoughts on geopolitical diversity was the exogenous interest in the Central Asian periphery with long-lasting implications for the livelihoods of people. The major impact seen until today is the delineation of international boundaries and internal borders. Most of the mountain region became an even greater periphery after border demarcation and lost its prime commercial asset as a transit region for traders. The deadlock situation has partly changed since the end of the Cold War, but not in a great style of regional cooperation. Regularly open borders are closed as the result of bilateral political disputes or so-called security issues. Territorial claims are not unanimously settled until today. Therefore the mountain periphery of Central Asia is not an acknowledged territory by international legal standards.

The second exogenous intervention had even greater impact especially on Kyrgyzstan and Tajikistan. About 75 years ago the major transformation of socio-economic conditions took place. The Soviet modernization project changed lifestyles and civil rights. To quote contemporary sources on the contents of Stalin's modernization project:

> "The CPC of the Tajik S.S.R. is drawing up a plan for agriculture in the Pamirs, the idea being thereby to transform the migrant tribes into stationary inhabitants, and to encourage them to grow their own food instead of importing it. A biological station on the Pamirs, at a height of 4,000 metres above sea-level, is just being started" (PRAVDA 7.5.1934, quoted after IOL/P&S/12/2273).

The "Pravda" told the truth: modernization meant the sedentarisation of nomads which was executed with great force and rigour. The effects of settlement and the introduction of "modern" animal husbandry can be observed in all areas north of the Amu Darya while on the southern bank of the river "traditional" forms of livestock-keeping prevail.

Similar developments could be observed in people's organization, education and agriculture. To quote again a source from 1934:

> "Khorog is the capital town of the Soviet Pamir, and there has been held there the 5th congress of the Soviets of the mountainous Badakhshan region. On foot on horses, on yaks, on donkeys, along mountain tracks hanging over precipices, the delegates come

from the distant Murghab, Borgang [Bartang], Bakhan [Wakhan], and other places in the S. and E. edges of the U.S.S.R. that border with Afghanistan, India and Western China. The 110 delegates elected were 78 Tajiks, 16 Kirghiz, and 16 Russians. In the conference hall were many women in their white garments of homespun silk. Khorog is now lit with electricity that was started and first seen by the Pamir people in the spring of this year. The president of the congress, Faisilbekov, spoke of the wonderful things that have taken place in the Soviet Pamir. Aeroplanes are flying over inaccessible mountain ranges, a splendid automobile road has been made from Khorog to Osh, 700 km long, that now links the Pamir with the rest of the U.S.S.R. formerly there was only 1 school in the whole of the Pamirs – now there are 140, and a training school for teachers: instead of dark smoky earth huts or skin tents, European houses are now being built: collective farms are established in the Pamirs, and they are growing and getting good crops of wheat, millet and beans; and now they know how to manure their fields and be sure of good crops" (Izvestiia 29.11.1934, quoted after IOL/P&S/12/2273).

It is the irony of history that now a transformation process has started which attempts to revert these reforms and to privatize collectivized property (Fig. 9) again, in which households return to the farming practices of their grandfathers, and in which the traditional knowledge of neighbouring countries is adapted as a measure to overcome food crisis situations and to minimize risks. In that respect the external interference in Central Asia is a failed attempt to implement modernization theory while in many other aspects it succeeded. The transition beginning with the independence of sovereign nation states in Central Asia has failed so far to continue the path of modernization.

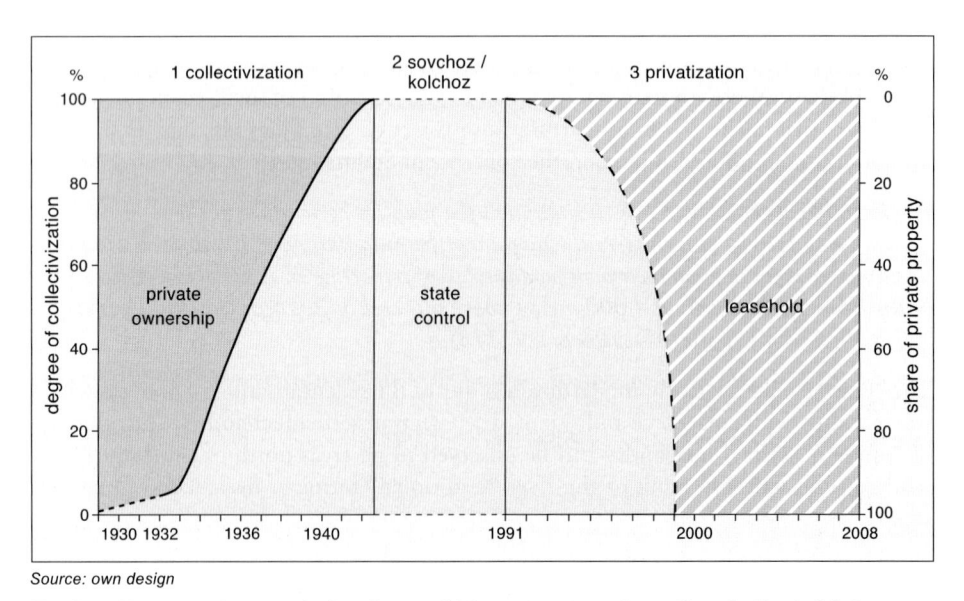

Source: own design

Fig. 9   Property changes during the two 20th century transformations in Central Asia

7 Conclusions

The lesson to be learnt from geopolitical interventions in peripheral mountain areas could be that decisions made in the core of empires always affect the livelihoods of people who have not been involved in the decision-making process. Socio-political interference led to the creation of an arena of confrontation in the Pamirs, Hindukush and Himalaya during the Cold War which was one of least permeable frontier regions in the world. Present developments might result in a convergence of living conditions, income patterns and indicators of human development. Especially mountain farmers and breeders can learn from the experiences of their counterparts, entrepreneurs might profit from trans-border exchanges in a way which was impossible for more than two generations.

Geopolitical diversity in mountain regions can be interpreted as the result of political decision-making in the centres of power. In a more system theoretical perspective it reflects the observation that changes in singular system elements are affecting the whole system. In the context of mountain development it often has detrimentally changed the living conditions. Nevertheless, the Amu Darya river boundary convincingly shows how life has been altered during two transformations in the 20th century. The transformation which is taking place in Afghanistan at the beginning of the 21st century takes up some of the threads which were loosened since the dissolution of the Soviet Union and the independence of Tajikistan. The Afghan National Solidarity Programme aims at establishing people's representation on the local and regional level. In Northern Afghanistan's history this is a new challenge and endeavour. Donor agencies support those activities which could lead to a bottom-up approach in communal participation. The same agencies are present on both banks of the Amu Darya River to mitigate economic and nutrition crises which are perceived by the local people as system crises. The understanding of which system has collapsed differs quite substantially. The same holds true for the path of development in a diverse mountain environment.

8 References

Bosshard, W. (1929): Politics and Trade in Central Asia. Journal of the Central Asian Society XVI, pp. 433–54.

Bregel, Y. (2003): An Historical Atlas of Central Asia. Leiden, Boston: Brill (Handbook of Oriental Studies, Section 8: Central Asia, Vol. 9).

Christian, D. (2000): Silk roads or steppe roads? The silk roads in world history. Journal of World History 11 (1): 1–26.

Davis, L. E.; R. M. Huttenback (1987): Mammon and the pursuit of Empire: The political economy of British Imperialism, 1860 – 1912. Cambridge: Cambridge University Press.

Djan-Zirakyar, R. R. (1978): Stammesgesellschaft, Nationalstaat und Irredentismus am Beispiel der Pashtunistanfrage. Frankfurt/Main.

Ehlers, E.; H. Kreutzmann (eds.) (2000): High mountain pastoralism in Northern Pakistan. Stuttgart: Franz Steiner-Verlag (= Erdkundliches Wissen 132).

Felmy, S. (1993): The Dividing Line. Newsline 5 (5–6): 72–78.

Felmy, S.; H. Kreutzmann (2004): Wakhan Woluswali in Badakhshan. Observations and reflections from Afghanistan's periphery. Erdkunde 58 (2): 97–117.

FRASER-TYTLER, W. K. (1953): Afghanistan. A study of political developments in Central and Southern Asia. London, New York, Toronto: Oxford University Press.

GILLS, B. K.; A. G. FRANK (1991): 5000 years of World System History: The Cumulation of Accumulation. In: CHASE-DUNN, C.; T. HALL (eds.): Precapitalist Core-Periphery Relations. Boulder: Westview Press: 66–111.

HALBACH, U. (1992): Ethno-territoriale Konflikte in der GUS (= Berichte des Bundesinstituts für ostwissenschaftliche und internationale Studien 31–1992). Köln: Selbstverlag.

HAUNER, M. (1989): Central Asian Geopolitics in the Last Hundred Years: A Critical Survey from Gorchakov to Gorbachev. Central Asian Survey 8: 1–19.

HAUNER, M. (1992): What is Asia to us? Russia's heartland yesterday and today. London: Routledge.

IOL/P&S/7/82, India Office Library & Records: Files relating to Indian states extracted from the Political and Secret Letters from India 1881 – 1911: Pamir Delimitation Commission 9. Oct. 1895 No. 195 (Reg. No. 451).

IOL/P&S/12/1321, India Office Library and Records: A survey of Anglo-Afghan Relations 1747 – 1947. Part III The Indo Afghan Frontier.

IOL/P&S/12/2273, India Office Library and Records: Departmental Papers: Central Asia. Conditions in Soviet Central Asia. Central Asia Intelligence 1930 – 1945.

IOL/P&S/12/2275, India Office Library and Records: Departmental Papers: Central Asia. Conditions in Soviet Central Asia. Central Asia Intelligence 1930 – 1945: Special survey of intelligence. Conditions in Central Asia and Sinkiang. 1939.

IOL/P&S/12/2354: India Office Library and Records: Departmental Papers: Political and Secret Internal Files and Collections 1931 – 1947: Captain K. C. Packman, Consul-General at Kashgar: Trade Report 1937.

IOL/P&S/18/C 17: India Office Library and Records: Political and Secret Department Memoranda: The Progress of Russia in Central Asia by Colonel M. J. Veniukoff (translated from the "Sbornik Gosudarstvennikh Znanii" [Collection of Governmental Knowledge] 1877.

International Crisis Group (2002): Central Asia: Border disputes and conflict potential. Osh, Brussels (= ICG Asia Report No. 33).

KHAN, Khushi M. (1981): Der Paschtunistan-Konflikt zwischen Aghanistan und Pakistan. In: KHAN, Khushi M.; V. MATTHIES (eds.): Regionalkonflikte in der Dritten Welt. Ursachen – Verlauf – Internationalisierung – Lösungsansätze. München: 283–385.

KREUTZMANN, H. (1996): Ethnizität im Entwicklungsprozeß. Die Wakhi in Hochasien. Berlin: Dietrich Reimer-Verlag.

KREUTZMANN, H. (1997): Vom Great Game zum Clash of Civilizations? Wahrnehmung und Wirkung von Imperialpolitik und Grenzziehungen in Zentralasien. Petermanns Geographische Mitteilungen 141 (3): 163–186.

KREUTZMANN, H. (1998): The Chitral Triangle: Rise and Decline of Trans-montane Central Asian Trade, 1895 – 1935. Asien-Afrika-Lateinamerika 26 (3): 289–327.

KREUTZMANN, H. (2002): „Great Game" in Zentralasien. Eine neue Runde im Grossen Spiel? Geographische Rundschau 54 (7–8): 47–51.

KREUTZMANN, H. (2003): Ethnic minorities and marginality in the Pamirian knot. Survival of Wakhi and Kirghiz in a harsh environment and global contexts. The Geographical Journal 169 (3): 215–235.

KREUTZMANN, H. (2004): Ellsworth Huntington and his perspective on Central Asia. Great Game experiences and their influence on development thought. GeoJournal 59: 27–31.

Kreutzmann, H. (2006): People and Mountains: Perspectives on the Human Dimension of Mountain Development. Global Environmental Research 10 (1): 49–61.

Kreutzmann, H. (2007): The Wakhi and Kirghiz in the Pamirian Knot. In: Brower, B.; B. R. Johnston (eds.): Disappearing peoples? Indigenous Groups and Ethnic Minorities in South and Central Asia. Walnut Creek: Leftcoast Press: 169–186.

Lattimore, O. (1950): Pivot of Asia. Sinkiang and the Inner Asian Frontiers of China and Russia. Boston: Little, Brown & Co.

Mackinder, H. J. (1904): The Geographical Pivot of History. The Geographical Journal 23 (4): 421–444.

Rashid, A. (2000): Taliban. Islam, Oil and the New Great Game in Central Asia. London, New York: Tauris.

Roy, O. (2000): The new Central Asia. The creation of nations. London, New York: Tauris.

Schetter, C. (2007): Talibanistan – Der Anti-Staat. Internationales Asienforum 38 (3–4): 233–257.

COLLOQUIUM GEOGRAPHICUM

Vorträge des Bonner Geographischen Kolloquiums
zum Gedächtnis an Ferdinand von Richthofen

*Lectures by Bonner Geographisches Kolloquium
in memory of Ferdinand von Richthofen*

Band 2:	CONZEN, M. R. G.: Geographie und Landesplanung in England. 1952. 83 S.	€ 3,00
Band 4:	WAIBEL, L.: Die europäische Kolonisation Südbrasiliens. Bearbeitet von G. Pfeifer. 1955. 152 S.	€ 4,00
Band 7:	PARDÉ, M.: Influences de la Perméabilité sur le Régime des Rivières. 1965. 100 S.	€ 6,60
Band 8:	BÜDEL, J.: Die Relieftypen der Flächenspülzone Süd-Indiens am Ostabfall Dekans gegen Madras. 1965. 100 S.	€ 7,40
Band 10:	LAUER, W., P. SCHÖLLER, G. AYMANS: Beiträge zur geographischen Japanforschung. 1969. 80 S.	€ 1,50
Band 12:	LAUER, W. (Hrsg.): Argumenta Geographica. Festschrift Carl Troll zum 70. Geburtstag. 1970. 295 S.	€ 14,00
Band 13:	LAUER, W. (Hrsg.): Klimatologische Studien in Mexiko und Nigeria. Beiträge zum Problem der Humidität und Aridität. 1978. 190 S.	€ 21,00
Band 14:	TERJUNG, W. H.: Process-Response Systems in Physical Geography. 1982. 65 S.	€ 8,00
Band 15:	AYMANS, G., H. J. BUCHHOLZ, G. THIEME (Hrsg.): Planen und Lebensqualität. 1982. 272 S.	€ 19,00
Band 16:	ERIKSEN, W. (Hrsg.): Studia Geographica. Festschrift Wilhelm Lauer zum 60. Geburtstag. 1983. 422 S.	€ 23,00
Band 17:	Richthofen-Gedächtnis-Kolloquium – 26.11.1979. 1983. 58 S.	€ 11,00
Band 18:	KEMPER, F.-J., H.-D. LAUX, G. THIEME (Hrsg.): Geographie als Sozialwissenschaft. Beiträge zu ausgewählten Problemen kulturgeographischer Forschung. Wolfgang Kuls zum 65. Geburtstag. 1985. 372 S.	€ 23,00
Band 19:	AYMANS, G., K.-A. BOESLER (Hrsg.): Beiträge zur empirischen Wirtschaftsgeographie. Festschrift Helmut Hahn zum 65. Geburtstag. 1986. 238 S.	€ 24,00
Band 20:	EHLERS, E. (Hrsg.): Philippson-Gedächtnis-Kolloquium – 13.11.1989. 1990. 95 S.	€ 17,00
Band 21:	BÖHM, H. (Hrsg.): Beiträge zur Geschichte der Geographie an der Universität Bonn. 1991. 423 S.	€ 26,00
Band 22:	EHLERS, E. (Hrsg.): Modelling the City – Cross-Cultural Perspectives. 1992. 132 S.	€ 23,00
Band 23:	GRAAFEN, R., W. TIETZE (Hrsg.): Raumwirksame Staatstätigkeit. Festschrift für Klaus-Achim Boesler zum 65. Geburtstag. 1997. 309 S.	€ 23,00
Band 24:	EHLERS, E (Hrsg.): Deutschland und Europa. Historische, politische und geographische Aspekte. Festschrift zum 51. Deutschen Geographentag Bonn 1997: „Europa in einer Welt im Wandel". 1997. 310 S.	€ 22,00
Band 25:	EHLERS, E (Hrsg.): Mensch und Umwelt. Gedanken aus Sicht der Rechtswissenschaften, Ethnologie, Geographie. Laudationes und Vorträge gehalten aus Anlass der Verabschiedung von Frau Ursula Far-Hollender. 2001. 71 S.	€ 8,00
Band 26:	WINIGER, M. (Hrsg.): Carl Troll: Zeitumstände und Forschungsperspektiven. Kolloquium im Gedenken an den 100. Geburtstag von Carl Troll. 2003. 120 S.	€ 15,00
Band 27:	RICHTER, S.: Wissenschaftliche Nachlässe im Archiv des Geographischen Instituts der Universität Bonn. Findbücher zu den Nachlässen von Carl Troll und Alfred Philippson. 2004. 556 S.	€ 29,00
Band 28:	LÖFFLER, J., U. STEINHARDT (Hrsg.): Landscape Ecology. 2007. 62 S.	€ 10,00
Band 29:	WIEGANDT, C.-C. (Hrsg.): Beiträge zum Festkolloquium aus Anlass der Benennung des Hörsaals des Geographischen Instituts in „Alfred-Philippson-Hörsaal". 2007. 139 S.	€ 18,00
Band 30	BURGGRAAFF, P., K.-D. KLEEFELD (Hrsg.): Entdeckungslandschaft unterer Niederrhein – Land zwischen Maas und Rhein. Neue Forschungen zur Kulturlandschaft des Niederrheins auf der Grundlage der Arbeiten von Gerhard Aymans, und Rudolf Straßer. 2008. 147 S.	€ 15,00

In Kommission bei • *on consignment by* Asgard-Verlag, Sankt Augustin

Nicht genannte Nummern sind vergriffen, sämtliche Titel unter
Titles not listed are out of print, see for all titles   www.geographie.uni-bonn.de/schriften.welcome.html